Digital Assistive Technology

A Guide for People with Disabilities

Edited by
Nick Awde,
David Banes &
Katharine Banes

MILLENNIUM
COMMUNITY SOLUTIONS

Contact us at:
info@millenniumcommunitysolutions.com
(+44) 07368 487263

First published
in 2022 by

MILLENNIUM COMMUNITY SOLUTIONS

&

DESERT♥HEARTS

www.deserthearts.com

© Millennium Community Solutions 2022

Typeset and designed by Desert♥Hearts

Printed and bound in Great Britain by
Biddles, King's Lynn, United Kingdom

British Library Cataloguing in Publication Data
A catalogue record for this book is available from the British Library

ISBN 9781908755544

*

Every effort has been made to contact and obtain
permission from owners of copyrighted material
included in this book. In case of oversight
please contact the publishers.

This book is sold subject to the condition that it shall not, by way of
trade or otherwise, be lent, resold, hired out or otherwise circulated
without the publisher's prior consent in any form of binding or cover
other than that in which it is published and without a similar
condition being imposed on the subsequent purchaser.

Contents

About Millennium Community Solutions	4
Foreword	7
Introduction to this guide	9
What is assistive technology?	10
Choosing assistive technology	12
For people with physical disabilities	15
For people with a visual impairment	21
For people with a hearing loss or who are Deaf	27
For people with neurodiversity and cognitive disabilities	31
For people with communication impairments	37
For people with mental health needs	42
In the workplace	45
In education	49
Daily living	54
Health	59
Leisure	63
Mobility	69
Assistive technology – a summary	72
Products – assistive and accessible technologies	74
Further sources of information	234
Product index by needs	235
Product index by tasks and settings	237
Products that are free	239

About Millennium Community Solutions

Millennium Community Solutions C.I.C. (MCS) was created in 2020 by Reverend Gail Thompson, to provide Assistive Technology (AT) resources such as specialist computer equipment, software and training – AT solutions tailored to your requirements and any configuration personalising the system.

Millennium Community Solutions has also started developing critical coding STEM skills initiatives for children, youth empowerment and development, recognising and nurturing the strengths, interests and abilities of young people through the provision of real opportunities that will meet employment and socio-economic needs.

We deliver the coding skills that help people switch to a career in software development within a year. Demand for software developers is rapidly escalating as digitisation accelerates, and hundreds of new roles open. Our online National Lottery-funded Skills for Tomorrow programme gives learners with a physical disability or those who are neurodivergent a direct pathway into these careers.

The AT Guide Team

Reverend Gail Thompson: "I am a British-born Caribbean who has lived with Multiple Sclerosis and I have been a wheelchair user for 34 years. I founded Millennium Community Solutions in 2021. Prior to that, my career spanned a range of ministerial and senior roles. It is a real privilege for me to be working with statutory bodies and stakeholders that seek to promote positive and progressive approaches towards inclusive education, and disability. Millennium Community Solutions is working to harness skills, knowledge, and collectively challenging barriers to shape the borough of Lambeth through social value."

Nick Awde is a journalist and cultural producer with a focus on inclusion and diversity in the arts. He is publisher and editor of Desert Hearts and co-director of the Alhambra Theatre (Alhambra Live) in Morecambe.

David Banes is Director of David Banes Access and Inclusion Services and co-founder of Global symbols. He worked as headteacher in a school for children with disabilities and has led assistive technology services in Europe and the Middle East. He now works across the globe, including projects with low and medium-income countries in Africa and South Asia to support the implementation of access and inclusion from policy to practice.

Katharine Banes taught English and Humanities for many years, including Special Needs Support services, and produced materials for European projects on communication and people with a disability. In addition, she works with children on personal safety and on research projects related to access and inclusion.

Isabel Appio is responsible for the day-to-day operations surrounding the Guide as well as building networks to connect us with other Assistive Technology companies and partners.

A Note on the information in this Guide

Every effort has been made to ensure that the information contained in this Guide is accurate, and details may be subject to change. If you have any questions, comments or suggested changes, please contact us at info@deserthearts.com

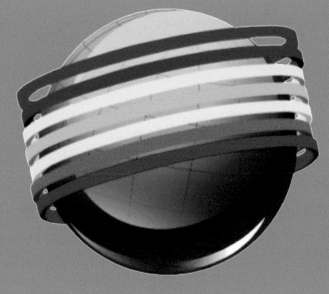

MILLENNIUM
COMMUNITY SOLUTIONS

Our Vision is to make information technology and education accessible for everyone, at home, school or in the workplace.

We strive to provide Assistive Technology tools and guidance that allows everyone to have a comfortable, and effective computer experience. And to prepare children & young adults to better equip them to compete in a global digital skills workforce through coding & STEM skills.

Get in touch with us by email or phone:

gail@millenniumcommunitysolutions.com

(+44) 07368 487263

millenniumcommunitysolutions.com

Foreword

by Emma Weston OBE
Chief Executive, Digital Unite

This is an extraordinary book that you have opened today. It truly has the power to change lives – maybe, hopefully – your life. That's because the digital world holds opportunity for all of us, and feeling comfortable and confident in it is something that should be possible for anyone and everyone.

Sadly, frustratingly, this is not always the case. As in so many areas of life, our own personal circumstances, constraints and restrictions can impact our capacity to engage digitally. Wider issues to do with where and how we live can also have an influence.

The paradox is that digital technology should be – can be – one of the most open, accessible spaces of endless possibility, for all of us, regardless of who we are and 'how we show up'.

Now that possibility is, quite literally, in your hands.

Millennium Community Solutions have brought together in this one extensive volume everything you need to know about how to make the digital world work for you and for others who may need extra support.

I would encourage you to start at the beginning and read both the Introduction and the following two pages that explain 'What is assistive technology' because they provide such a clear starting point for your journey into the detail. Because, as every page testifies, that journey really IS detailed. And it's thrilling! There are ways to support and ease the digital experience for everyone, including things that 'mainstream' digital users would be well served to understand as well.

It's in everyone's interests that everyone has equal access to digital technologies and the skills and confidence to use them. Writing this Foreword in the autumn of 2022 as the UK plunges into a cost-of-living crisis and with winter ahead, we are all to greater and lesser degrees braced. Being digitally capable is not a panacea, it's not a silver bullet – the world will not turn on its axis reconstituted because more of us are confidently accessing the internet using a range of devices adapted to our individual needs.

But, being digitally included will unquestionably help us all – to access information, stay connected to friends and family, find cheaper deals on anything under the sun from buying food to clothes to holidays. It will support us to claim benefits and support, to request and access health and social care services. It will support us in our studies, to get skills and to find, and then manage, work.

In producing this brilliant Guide, Millennium and their partners and collaborators are making all of that more possible for more people. I congratulate and thank them for such a thorough piece of work. And you, dear Reader, for picking it up. Here begins your supported discovery of new digital opportunities that I hope will benefit you for years to come, and which you may in turn share and encourage others to explore also.

*

Introduction to this guide

Millennium Community Solutions is a community interest company based in South London. Part of our work focuses on ensuring that people living with disabilities are valued as important contributors to society. Central to this is the right of people living with disability to have equal access to training and technology to support day-to-day life, employment and social connectivity alongside dignity and respect for their circumstances and needs.

This book aims to provide a toolkit to help guide anyone with a disability in choosing assistive technologies (AT) that meet their needs. In preparing the guide, we felt it was important not only to raise awareness of the many available options but also to help build the capacity of those with disabilities to make informed decisions wherever possible. As a result, the book suggests a wide range of technologies to help address challenges people with various needs face. We also seek to suggest technologies available at very low cost or are free so that people can start to try out solutions as soon as possible.

In preparing this guide, we have recognised that the technology that benefits many people is not always specially designed to address their needs. Instead, consumer or 'mainstream' technologies often have features or functions which have the potential to make a significant difference in lives. So as you work through each chapter, you will find constant reference to devices, aids and gadgets that can be easily purchased in the high street or through an online website or app. We often recommend that these technologies be a great starting point when reducing barriers in everyday living, drawing upon the vast experience of people with disabilities in using and testing technology.

This guide to assistive technology is divided into seventeen chapters. The first two chapters introduce the concept of assistive and accessible technologies. Helping us understand that both 'types of technology' are valuable. We will also talk a little about some of the questions you should think about when choosing tech you would want to try. The next six chapters explore technologies available to meet needs in different settings. Whilst some forms of assistive technology are suitable across all parts of our lives, others are more specific to one or more parts – e.g. for education, leisure or at work. The final two chapters summarise everything that has been discussed, while also providing you with additional sources of information and support to move forward and start using technologies to make life easier.

The book draws upon the experience of many people. Their insight at conferences and exhibitions, in discussions and workshops was a crucial influence on the advice contained in this toolkit. The lived experience of people with a disability – as expressed on social media, at events, and in the public feedback they give for products – was a vital source of information. We thank everyone for their willingness to share that experience.

what is AT?

What is assistive technology?

Assistive Technology is defined by the Assistive Technology Industry Association as "any item, piece of equipment, software program or product system that is used to increase, maintain or improve the functional capabilities of persons with disabilities". AT can be much more – electronic devices, wheelchairs, walkers, braces, educational software, power lifts, pencil holders, eye-gaze and head trackers, and much more.

AT can be:

- Low-tech – communication boards made of cardboard or fuzzy felt
- High-tech – special-purpose computers
- Hardware – prosthetics, mounting systems, and positioning devices
- Computer hardware – special switches, keyboards, and pointing devices
- Computer software – screen readers and communication programs
- Inclusive or specialised learning materials and curriculum aids
- Specialised curricular software

Assistive technology helps people who have difficulty speaking, typing, writing, remembering, pointing, seeing, hearing, learning, walking, and many other things. Different disabilities require different assistive technologies. It is perhaps possible for people with a disability to think about it as 'technology that helps us' – although this is not a definition that would fit most funding.

Within this toolkit, we make more specific references to three varieties of 'technology that helps us'.

Mainstream consumer technology

These technologies, such as digital television or smart doorbells, are designed for use by the broadest population. Therefore they may include no specific features to facilitate their ease of use by a person with a disability. For example, a TV may have Alexa built in, but may not enable access to captions for hearing loss or audio descriptions for the blind. Still, the large screen and control may work well for some people with some form of disability. Featurephones may be examples of such technology with relevance to our discussion.

Accessible technologies

Accessible technology includes products, equipment, and systems that can be customised and provide persons with disabilities access to all services and content. Some examples might include smartphones with integrated speech-to-text that can be used for dictation or creating captions. Such devices are widely used. The features may benefit anyone using the equipment and are designed to be used with minimal effort to meet the needs of a broad population.

Assistive technologies

'Assistive technology' refers to products, equipment, and systems that enhance learning, working and daily living specifically to address the needs of persons with disabilities. However, many assistive technologies are unlikely to be required for most people.

Such technologies can include screen readers, braille output, connections to hearing aids and alternative access technologies for those with a physical disability, including pointing devices or switch access. In addition, the World Health Organisation's 2016 priority list of assistive products includes devices such as braille notetakers. The World Health Organisation and its global initiative GATE have sought to clarify some of the terminology and concepts.

GATE defines 'assistive technology' as a wider concept. This includes both systems and services alongside products. The GATE definition emphasises assistive technology as enhancing functioning, not simply as medical or health interventions. It continues to refer to assistive technologies as a subset of health technology, which may be due to the legacy of the full breadth of assistive technology and products, including wheelchairs, mobility aids and prosthetics.

Much of the functionality of what is described as assistive technology can now be delivered through the integrated features or additional apps, on a smartphone or similar device.

The importance of mobile and consumer technology

Mobile and consumer technologies for many people with a disability, which may or may not have additional assistive technologies to add on, are critical to ensuring independence. Most people with a disability are not concerned about whether the technology is for 'everyone' or has been specially designed for them. Instead, they simply want to have technology available to them to make a difference in their lives and reduce barriers.

As a result, we have explored how features incorporated into a device make a difference in the toolkit. This may be to make the device accessible or offer a way of fixing other real-world challenges.

But, as there may be other ways of purchasing the equipment that would be helpful, we have tried to include as many options as possible to guide decision-making. But, the more a device addresses barriers to inclusion, the more likely it can be defined as assistive technology. For instance, JAWS is often funded as an assistive technology because it creates access to a computer but is not part of the computer when supplied. It is an independent product that addresses problems in the design of computers for people with little or no vision. However, the Narrator is not an assistive technology because it is integrated into the device and embedded to make it inclusive.

Ultimately, our criteria for including products in this toolkit are less about standards and classifications and more about whether this technology will help people.

Choosing assistive technology

With so many possible technologies to choose from, it can feel daunting to decide. Traditionally, many people with a disability turned to professional services to help them choose. This was important when the technology was expensive. It was impossible to get funding to pay for the device or solution without a professional recommendation. But the world has a terrible habit of changing, and sometimes it feels as if nothing remains constant. The world is far more chaotic than we would wish for. Change has driven the availability of assistive technology services for several years, and the pace and diversity of change have probably never been greater. We face a changing demographic that has increased demand for products, economic pressures upon the public purse and more people with a disability voicing their aspirations and frustrations. Those services and the provision of products can feel as if they are being built upon shifting sands, with increased expectations of on-demand services reflecting broader social and consumer trends.

We hesitate to mention the Covid pandemic at this stage. The sudden injection of enormous stress into systems has challenged much of our thinking, and assumptions about what is 'best practice' vital services have collapsed. But even before this, social trends were creating new demands upon systems. These, combined with economic pressure and technology innovation, has led us to rethink how to find and choose the technology we want and need. These pressures have brought growing recognition of the value of embracing greater self-determination of assistive technologies by people with a disability themselves. This shift has implications for the funding of assistive technology and the role of assistive technology professionals.

Self-determination of assistive technologies is not a new phenomenon. Many assistive technologies have always been made available through consumer choice. For example, reading glasses, simple hearing devices, manual wheelchairs, crutches, and aids for self-care are often made available to people with a disability through high street stores. In these stores, 'try and buy' is at the heart of the decision-making process. However, with the advent of digital assistive technologies, embedded or added to consumer devices and replicating the functionality of dedicated aids, there has been a transformation of access to the tools that people with a disability require.

In developing this guide, we reflected upon the confidence felt by people with a disability in a series of workshops held before the pandemic. We used the SETT concept developed by Joy Zabala as a starting point. SETT consists of four concepts: Student (or Person) needs, Environment, Tasks and Technology. Technology knowledge was where people with disabilities felt they needed the greatest support. The other three dimensions, relating to their needs, the setting in which the assistive technology was to be used, and the tasks or activities they wanted to undertake were all areas they felt confident to describe.

In many cases, the try and buy approach worked well. They could handle or even trial products off the shelf of a store before making a final decision. Then, with some advice from salespeople, the decision rested firmly in their own hands. With the advent of assistive and accessible technologies integrated with a mobile phone or tablet or available as consumer technology, the same fundamental approach can be applied. If we want to use our phone as a handheld magnifier, we can download and evaluate at least six free or low-cost apps. If we want to have directions spoken out, mainstream apps offer voice navigation. We can try and buy software to help us at home, at work or at school or college. Try, Test, Choose and Buy becomes a simple route by which decisions can be made. Professionals add considerable value by helping the user find trusted information from an independent perspective on what is possible to start that process of making choices.

Using SETT as a guide to choosing

The SETT framework for choosing assistive technology has been successfully implemented for many years. Joy Zabala, the creator of the framework, states on her website (www.joyzabala.com), the framework, is a four-part model intended to promote decision-making by considering the following questions.

What are my needs?

Here we focus is the person. We might want to think about:

- What functions cause me problems? These could include reading, personal organisation, mobility, dexterity, communication or sensory loss.
- What is it that I struggle to do independently?
- How would I describe my current needs, expectations, interests, and preferences?
- What money do I have to pay for my technology?
- How confident am I with technology?
- Are there any language issues to consider?

These leading questions might prompt us to think hard about the specific areas of disability that we want to address.

Where do I want to use this technology?

In this section, where 'E' stands for 'Environment', we try to concentrate on where the activity will take place and what influence this might have on choices.

We might want to think about:

- How much control over the physical space do I have to use this technology?
- Is space an issue or are there issues with access to power, and will available lighting create any problems?
- Am I sharing the space with other people? Will my technology cause any problems for them?
- Will I have access to the internet?
- What other technology will I need to use with my assistive technology, applications, digital content such as ebooks and websites or other devices?
- Is there any help available to me when things go wrong? (They will.)

# choosing assistive technology

- Are there any other access issues for me in this environment like stairs, doorways, noise?
- What are the attitudes and expectations of the team who might be available to help me? For instance, the IT team at school, college or the workplace.

What tasks do I need to complete?

This section of the framework asks us to consider the specific tasks we need to undertake:

- In the workplace, this might be based upon our job description.
- In education, it is related to the curriculum.
- In everyday life, these may relate not only to how we care for ourselves but might also be concerned with our health, social life and leisure pursuits, personal mobility, travel or transport.
- We might want to address tasks from more than one setting.

What technology options are available to me?

Finally, we start to think about technology and tools. As we will discover in this toolkit, these might include devices, software, apps and services. The technologies available might address our specific needs across settings or focus upon issues in one setting, such as a classroom. They might help us succeed in one or many tasks and be very specialised or mainstream.

It can be useful to think about how much impact we think this technology will have in making choices. Sometimes this can be very broad across every aspect of our lives, and sometimes very deep, fixing one specific problem that otherwise cannot be resolved. Both are fine!

*

If technology is cheap and low cost, don't be scared to borrow, trial, download, and share products. It is often the fastest way of deciding what you want. But if this turns out not to be enough, never be afraid to turn to other sources of advice. Those may be informal on social media or more formally through assistive technology services.

For people with physical disabilities

People with a physical disability face many challenges in daily life. Continued access to independence and personal mobility remains a significant barrier for many. Often the means of addressing this is through the availability of wheelchairs or other types of mobility aid, from crutches to walking aids. Mobility needs may not create any additional barrier for access to technology, although physical access to office space and appropriately designed and ergonomic desk space may be essential to both productivity and comfort.

Other people with physical needs face challenges characterised by hand/arm issues. These impact their physical dexterity and ability to manipulate and control technology through a mouse or standard keyboard or successfully use touch devices such as smartphones or tablets. For some people with a physical disability, the need will change with time. For instance, some people with a condition such as Multiple Sclerosis may find that they can sometimes cope with a keyboard and a mouse. Still, others find that software such as voice recognition is more suitable. In choosing a solution, many people must think about what works well for when they feel at their best and what they need when they find the challenges greater.

Some people with other conditions may also experience challenges with physical access, especially with hand/eye coordination. For example, as children, some may have struggled with ball games and, as an adult, still meet challenges due to a need for fine motor skills and dexterity.

The tools and techniques in this chapter help reduce the impact of such barriers to using technology by making the technology itself more accessible and helping with writing or using any available software or applications.

Options for making a device easier to use for people with physical disabilities

Simple Hardware additions

A keyboard is supplied with most computers, but the 'one size fits all' approach may not suit everyone. There are many types of alternative keyboards available, including those with high colour contrast keys for people with low vision and others with alternative layouts such as 'DVORAK', which are based upon the frequency of use of the keys and letters. These can work very well for some people where the standard layout is hard or uncomfortable. Other keyboards can be useful for those experiencing mild or moderate physical impairments. For example, compact keyboards are smaller keyboards like those found on a laptop. Its size allows it to be positioned easily on a desk or wheelchair tray. In addition, the small size and missing numeric keypad mean that any pointing device can be kept closer to the keyboard, which may make it easier to control.

Places

Though our focus at Millennium Community Solutions is on Assistive Technology, we are dedicated to re-opening and running spaces in our local community. Our first spaces are Tenants & Residents Association (TRA) halls in the Borough of Lambeth, South London. Through Lambeth London Borough Council and other stakeholders, we have been able to start reintroducing the following spaces to the local communities that so desperately need them.

Southwyck House Community Hall

Millennium Community Solutions secured the lease for Southwyck Community Hall in early 2020. Due to fantastic support from Lambeth London Borough Council and with help from T Brown, Derwent London, Breyer Foundation and OCO, we were able to completely remodel and re-purpose this wonderful space at the base of the beautiful Southwyck House Estate.

Millennium Community Solutions has been joined by Cre8tive Arts, generating a new cultural hub in the area and giving local artists an opportunity to have their work hosted in a top-of-the-range gallery space.

Southwyck Coding and Arts Centre (SCAC)

Southwyck House was built as a 'barrier estate' back when there were plans to extend the South Circular and build a Hammersmith-style motorway flyover above the residents of Brixton. The concept was that the building would protect other areas from the noise and air pollution of the soon-to-be constructed motorway. Thankfully construction never started and so we are left with this stunning example of Brutalist architecture from the 70s and 80s.

The Jubilee Centre

The Jubilee Centre was opened in October 2021 to serve the people of Norwood, especially those living on the Portobello Estate, Knights Hill. It was fully refurbished over the course of the year by our partners at T Brown and Wolseley Paints

The Jubilee Centre Community projects delivered by Bright Insight Ideas CIC, and activities will include welfare support, technology access Including job search/filling in government forms, occasional distribution of food parcels, workshops on health and well-being, as well as a safe place for young people of the surrounding areas. The centre is planned to be fully operational in 2022.

millenniumcommunitysolutions.com

A good example is the **Cherry** which can have a matching **Keyguard** attached for those that need it. Alternatively, others find a larger keyboard such as the **Clevy Keyboard** helpful. This is a keyboard with a standard layout but with bigger keys that are very clearly labelled. Like the Cherry, it is possible to attach a keyguard to the Clevy. A **keyguard** is a plastic or metal grid matched to a specific keyboard. They are made in various styles and sizes. The keyguard rests on top of the keyboard, and you press the keys through holes above each key. Anyone with poor motor control, including tremors, can press a key without accidentally pressing others. They are also useful for people who tire when typing, as they can rest their hands on the keyguard without pressing any keys. Alternatives to a regular mouse can also be helpful. These can be connected to computers and a phone or tablet where an onscreen keyboard or touch is difficult to use. Options include:

Ergonomic mice

These are mice with different shapes, sizes and weights that are usually much more comfortable to use.

Trackballs/rollerballs

These options are often easier for someone with hand/arm difficulties. This is because they don't require the same levels of accuracy with finger, hand, and wrist movements. In addition, in many cases, the mouse button is separate from the ball's movement, making it easier for a person to hold the cursor over items whilst clicking.

Joysticks

Joysticks come in many different sizes and styles. Almost all require less control than a trackball, and often different tops or grips can be fitted to suit different needs.

Touchpads

A touchpad refers to devices similar to the small mouse pad found in front of the keyboard on a laptop. The smaller movements required to operate these may be easier than a mouse for some people.

These devices work well when the computer is 'tuned' a little to use them. Most of the functions that help us to make these adjustments are found within the operating system and are often referred to as Ease of Access or the Accessibility options.

Switch access

Switches are on/off devices that can send a signal to a device. When combined with an interface that allows the device to recognise the signal and software on the device itself, they are a very cheap and reliable way to control technology. In addition, some switches connect to devices via a simple cable or Bluetooth and do not need an extra interface box.

Switches can be used with almost any consistent body movement. For example, they can be activated with a hand, head, foot or chin movement, and even by breathing in and out with

physical disabilities

a **sip/puff** switch. Choosing the right switch for you can be challenging. New switches such as the **Cosmo** with additional features are always entering the market.

Adjustments to your device

Ease of Access Centre/Accessibility Options

The accessibility options or ease of access centre are features that are built into most operating systems. They include features that make using a keyboard, mouse or alternative device easier. Some of the most common ones for Windows include Filter Keys, Sticky Keys and Mouse Keys.

When activated, Filter Keys prevents brief and repeated keystrokes from being registered as a keypress. Sticky Keys allows you to press one key at a time for keyboard shortcuts, and Mouse Keys allows you to use a numeric keypad to move the mouse around the screen. All are useful for those with a physical impairment and can be used with an alternative keyboard. Similar features are available within the operating system for phones and tablets.

Assistive Touch

A feature called Assistive Touch for iPhone and iPad allows the user to adjust volume, lock the screen, use multi-finger gestures, restart the device or replace pressing buttons with just a single tap. When activated, a button appears on the screen. This can be moved to any edge, and by tapping the button, a menu is opened. This is closed by tapping anywhere outside of it.

Assistive Touch can be used instead of gestures to access menus and controls that require onscreen gestures such as the control centre, notification centre or the app switcher. It can also be used as an alternative to pressing physical buttons such as activating the accessibility shortcut, adjusting volume, restarting the device or capturing a screenshot. It can also be helpful by simulating multi-touch gestures.

Alternative Home Screens

Whilst most smartphones and tablets have been developed to offer an interface that is as intuitive as possible; the design may still be overly complex and difficult for many with physical needs. As a result, alternative home screens have been developed which can be used instead of the standard screen. One good example for Android Phones is **Big Launcher**. The app replaces the home screens of most Android phones or tablets with enlarged buttons and texts. It has been designed for people with limited technical capacity to provide maximum readability and easy use. It is controlled by single touches and can be customised to match needs, putting shortcuts for apps, websites, contacts, widgets, and more directly on the home screen.

Similarly, **Project Ray** offers alternative front ends to Android phones, including Ray Vision. This allows users to change their phone's contrast and colour settings to achieve the most useful and personal level of accessibility. It also includes one-hand operation, text to speech, voice-operated speed dial, messaging services, call history, contact list, calendar, clock, voice SMS, social messaging, WhatsApp application, visual recognition, OCR capabilities and a lot more.

Extra ways to help with physical needs

Speech recognition

All popular devices have some form of voice recognition integrated into their operating systems and applications. For example, in Microsoft Office, a Dictate button is usually next to the 'editor' on the Home tab of the ribbon. In Google Docs, 'voice typing' can be found on the dropdown list from Tools on the main toolbar. Voice recognition is powerful but may not be suitable for everyone. A headset or high-quality microphone is recommended for computers, and anyone using it is strongly advised to check their text carefully before completing and sharing a document.

Dragon Naturally Speaking

Dragon Naturally speaking is a professional voice recognition package for computers. Whilst it is similar to the technology integrated into Office and Google docs, it has additional features. Some users find that it is more accurate in recognising their voices.

Word prediction

Word prediction packages offer more advanced text and sentence prediction, often learning from the text the author has written previously. Many people with physical disabilities write text more slowly than others, and text and sentence prediction can speed this up considerably. Whilst most technologies offer some word prediction, often as part of an onscreen keyboard, it can be beneficial to have the additional features of commercial software such as **Clicker, Text Help Read and Write** and **Lightkey**.

Head tracking

Head tracking is another way to control a pointer and cursor on a screen. Headtrackers usually operate through a camera that tracks the head movements of the user. Whilst they may be specialised assistive technologies such as **glassouse,** which is based upon a wearable headset, there are other options available that are widely used by gamers, such as **TrackIR** and **DelanClip**.

Eye tracking

Whilst often expensive, eye tracking devices such as the **Tobii, Skyle** or **Smartbox** systems are becoming increasingly popular to provide access for those with very limited control over their bodies. In most cases, the systems work through specialised cameras that focus upon the eyes and move a pointer on the screen as the user moves their eyes. By hovering over a link or cell on the screen, the device imitates the click of a mouse. The technology has proven especially powerful for people with motor neurone disease when control over one's body decreases as the condition progresses.

 physical disabilities

Consumer technologies

Customised phones and tablets

As adaptations to mainstream phones, several phones and tablets are designed to be more easily used by people with a disability. These often include bigger keys, simpler screens and easy to access voice control. For example, many people with tremors and limited dexterity use phones such as Doro and Emporia. More recently, companies have begun to sell tablets that come with accessibility features and apps preconfigured to make them easier to use. Many older people find these especially useful. While they may not have all the functionality of a modern smartphone, these may not be the features they would most require.

Amazon Echo and smart speakers

The rapid growth of smart speakers such as **Amazon Echo** and those from Google and Apple have helped to facilitate access for people with physical disabilities. Smart speakers can be used by speaking, asking for information or contacting someone. They can also connect to other devices to control the home or help to monitor health and well-being. As these are consumer technologies, they tend to offer a very low price point as an entry-level.

Ebook readers

Ebook readers have grown in popularity in the last five years, led by the success of the **Amazon Kindle**. Such readers can be hardware or software and offer access to millions of books online. The hardware readers can be used with a touch screen, whilst software for computers, phones and tablets can be used in combination with any other access method you might be using. The readers are especially useful for people with physical disabilities who may find turning pages difficult or carrying many books a problem.

*

People with physical disabilities, regardless of the cause of the disability, face many challenges. Ensuring that the available technology is accessible, usable, and comfortable for them is important for success. Product information is included in the directory of this book and the chapter on other sources of support.

visual impairment

For people with a visual impairment

In the case of those with a visual impairment, it is useful to understand that most people with such a need have some residual vision. However, this may have limited functionality as be limited to the ability to discern light or shapes. Therefore, we introduce assistive technology for people with visual impairment. Our aim is to maximise the use of any residual ability or offer access to information and technology in an alternative format where such ability cannot easily provide a firm foundation to build upon. Some users can be defined as 'blind' and, as a result, require technology that presents information and the user interface in an alternative manner. These are most likely to be a tactile form such as braille or an auditory form such as a screen reader.

Other users have a higher degree of residual vision. In these cases, those with a disability are most likely to benefit from the information presented on screen being magnified or offered with high contrast settings. We recognise that for people with some conditions, there will be times when your vision is better than on other days. So, it is often helpful to look at products for low vision and those with very limited vision in these situations. Equally, some products combine both speech output with magnification, which may be helpful. Finally, of course, you might use the accessibility features of the device as extra support.

Accessibility features of devices

LOW VISION

Windows Magnifier

Onscreen magnification is an important tool to support low vision, and Windows Magnifier has improved hugely in recent years. However, we sometimes get asked why not just increase font sizes in documents. Many people do this, but you increase that for every reader when you increase font size. In contrast, by magnifying your screen, that larger font is only seen by you. You will also find some extra 'zoom' features in Microsoft office products and other tools with the same impact, but only for one application and not everything on the computer.

Windows Magnifier can be activated in the ease of access centre in windows. It is a helpful starting point for many people when vision starts to deteriorate, especially with ageing. There are other simpler functions in the ease of access centre that can help make a pointer larger or easier to find. This can sometimes help resolve a very specific problem that many people experience.

Windows Contrast settings

Similarly, contrast settings in Windows allow you to change the colours used on your screen. Just as with Magnifier and Zoom, this only affects the document you see and not the document others receive. Many people with limited vision find that changing the colours and contrasts, such as white text on a black background or yellow text on a blue background, can make the text much easier to read.

visual impairment

NO VISION

Windows Narrator

If you need more help than the magnifier can offer, there is a simple screen reader built into Windows devices called 'narrator'. Whilst not as powerful as free and commercial screen readers, it is a useful tool. For those with low vision that varies, it can be especially helpful when your eyes become tired and sore. Sitting back and listening to a document or email can reduce that extra eyestrain and discomfort.

VoiceOver

Similarly, Apple devices, including iPhones and iPads, have their own screen reader called **VoiceOver,** which is well regarded by many people with both low vision and those who are blind. It offers a set of specific gestures to control the computer and speak out most things onscreen. VoiceOver also has support for a selection of **refreshable braille devices**. If you connect these to your Apple device, the screen content is represented in braille.

Talkback

Like VoiceOver for Apple, Android devices have a similar tool called **TalkBack**. This can be combined with **BrailleBack and KickBack** to offer non-visual notifications and refreshable braille support for those with little or no sight.

Assistive Technologies

Many people experience sight loss with ageing. They may find that glasses are insufficient for their needs, and there are many products to support low vision are on the market.

LOW VISION

Hi-Vis keyboards

Hi-visibility keyboards and keyboard stickers are useful for making the keys on a keyboard easier to find. Several companies produce specialised keyboards with black letters on a yellow background (or vice versa). As a low-cost solution, using coloured stickers that can be applied to your standard keyboard is a simple and useful solution.

Handheld magnifiers

Like a traditional magnifying glass, handheld electronic magnifiers are widely used to help read a fairly small amount of text such as labels, paragraphs of instructions and recipes. However, unlike a traditional lens, the amount of magnification is variable. In addition, electronic devices often include options to change colours and contrast what is being read, take a photo and enlarge it. This is especially helpful if the reader's hand is a little unsteady.

visual impairment

Closed circuit TV magnification (CCTV)

Larger versions of some of the magnifiers are also available, connecting a camera to a screen to view a whole document or page of a book. These usually have a standalone camera and a stand on which the book or page can be placed. The reader can then slowly move the page under the camera while reading the attached screen's enlarged and adapted text.

These devices have tended to be quite expensive in the past. Still, some lower-cost options are becoming available that use a computer or tablet with a separate document camera to offer some of the same functionality.

Magnification for phones and tablets

There are many applications designed for a mobile phone that replicate the functions of handheld magnifiers. One popular magnifier is **ClaroMagx** which allows for magnification, photographed text, and contrast settings in a free app. There are several other such applications available. Some are designed for people with low vision, whilst others might add the use of the camera light on a phone, mostly for people in low light settings who need to look at small objects or text. There are so many options, mostly available for free or less than one pound, but they are worth trying. Choose the one that suits you best, and then delete those you no longer need.

Some of the popular applications included on App Store and Google Play are: **Magnifying glass with Light** and **Seeing Assistant Magnifier**.

Commercial software for Windows

There are several commercial packages for those needing software with extra features and functions that also help bridge between using magnification and a screen reader. **ZoomText** is a magnifying package that includes speech. It offers high-quality magnification with extra features to make navigation around the screen easier. This has recently joined with the screen reader **JAWS** to offer a combined package that has been branded as Fusion. Another popular and widely used package that can combine magnification with a screenreader is **SuperNova.** SuperNova is available in three versions, magnification, magnification + speech, and magnification + a screenreader. It also includes the ability to recognise documents from a scanner or document camera such as the **Ziggi.** Supernova can also connect to a braille display as well as speaking the screen

NO VISION

There are also screenreader packages available for those with little or no vision other than SuperNova. These vary in price but offer much more powerful features than Narrator. The most widely used and free screenreader is **NVDA**, or Non-Visual Desktop Assistant. It was developed to be an affordable, free and fast solution for people with little or no sight, and is 'open source', which means that it can be distributed freely and has been translated into more than fifty languages. Like some commercially available packages, it supports **Refreshable Braille Displays**. Probably the most widely recognised screen reader is **JAWS**, or Job Access with Speech, which has been in widespread use for many years. It offers powerful features including the ability to customise or 'script' the software with extra lines of code.

visual impairment

JAWS supports a range of braille devices and is available in different versions for home, school or employment.

Refreshable braille displays

These are hardware devices made up of a row of braille 'cells' in a line. Each cell can represent a braille letter, and as a screen reader 'reads' the text onscreen, the characters change to make up words and sentences. They offer a way of accessing digital documents without printing out the whole document and carrying very bulky braille papers with you. Unfortunately, these devices are often very expensive, and most come as either 40 or 80 character displays. Most people who use a refreshable braille display find that 40 characters are sufficient.

Braille Notetakers are braille displays that also have a keyboard built in. This is usually a 'chording keyboard' where the typist presses combinations of keys to type a character. This can make typing faster. These devices are expensive and may cost more than £3000. But in recent years, new types of notetakers and displays have been brought to the market for much less cost. The **Orbit Notetaker** is a good example of a notetaker with similar features to the more expensive models that would be worth trying before deciding. Many people with vision loss also use apps and products that seek to address a specific barrier or challenge.

Be My Eyes is an app that connects a person who is blind with a sighted 'buddy' through a mobile phone. The volunteer uses the camera on the phone to view the world around the person with limited vision and describes what they can see to help solve a problem. This can be used in many situations from shopping to travel, or even simply when needing to sort coloured clothes from whites when washing them. **Seeing AI** is an app from Microsoft that uses artificial intelligence to recognise text, objects, labels and people and describe them. New features are always being added, including scene descriptions of the world around us. Wayfinding and orientation are important to people with vision loss to help them to travel independently. Speaking step by step directions are available in mainstream products such as Google Maps or Waze. But there are also more specialised wayfinding and navigation systems for people who are blind such as **Access Now**, **Aridane GPS** and **Blindsquare**.

Consumer Technology

Mainstream and consumer technologies are also vital in making the world a more accessible place for people who are blind or have low vision. These include wearables and smart speakers, tactile indicators, and speech recognition to increase access. For instance, **Siri** on Apple devices allows blind people to dictate text and control their phone or tablet just by using their voice. Similarly, **Amazon Echo** can respond to speech and provide audible news, books or entertainment through a low-cost device. Wearable devices such as the **Apple Watch** can also add haptic or vibrating alerts to the experience to notify someone that they have a message, email or update to listen to. The watch can also offer a simple way of accessing those messages, which can then be listened to using earphones connected to your phone.

Accessible content

People who are blind or have vision loss also need accessible content for use on their technologies. Traditionally many people subscribed to talking books services that would post human narrated recordings of newspapers and books to people who were blind. Producing these was time-consuming and expensive, and today, much of this work is done by technology.

Most books and papers can be scanned, and the words are turned into digital text using an OCR package such as OmniPage. But it is also possible to get access to hundreds of thousands of accessible books via mainstream suppliers such as Amazon for **Kindle** or specialist services such as **Bookshare.** Many of the books on Bookshare are available in braille-ready format, which can be printed or read on a braille display, whilst others are available in 'Daisy' format. Daisy is a talking book format like an ePub document that can be played back on a hardware layer such as the **Victor Reader Stream** or downloaded onto a phone and played through an app such as **Dolphin EasyReader**.

Other sources of spoken content include **Audible,** which has many audiobooks for download. These are human narrated on many occasions which some people prefer synthesised speech. Similarly, podcasts and radio remain popular sources of content. **Qcast** is a fully accessible podcast player. Although mainstream apps for apple and Android are often fully accessible, including radio and music apps such as Spotify and BBC Sounds.

*

There are many ways in which technology can benefit people who are blind or have low vision. There will be built-in features to make it easier to use with whichever device you have. Many add-on features and software applications can make the technology more accessible or use the device's power to address other challenges. Increasingly digital content is widely available in accessible formats from both specialised and mainstream sources.

Together we can end digital poverty – join us

digital poverty alliance

How you can help

We work with corporate and public sector partners to deliver innovative projects that close the digital divide, third sector partners to explore policy solutions, and research institutions to build the evidence base.

Join the community today:
https://bit.ly/DPAHub

Join us

 @DigiPovAlliance   digital-poverty-alliance

www.digitalpovertyalliance.org

digital poverty alliance

For people with a hearing loss or who are Deaf

There is a similar continuum of needs to that described in the chapter on vision impairment. These may be divided into Deaf, with little or no functional hearing, and those who are hard of hearing and may experience barriers to processing audible information depending on volume, clarity and distractions. Those who are hard of hearing often benefit from solutions that include amplification and noise reduction, whilst those who are Deaf may need support to sign language and captions. Although many of those described as hard of hearing also benefit from the availability of captions.

Current technologies for the Deaf and hearing-impaired fall into three broad categories: hearing, alerting, and communicating. Some forms of assistive technology seek to enhance hearing ability through volume and other sound adjustments such as noise reduction. These include hearing aids, assistive listening devices, personal sound amplification products, and cochlear implants. Cochlear implants are surgically implanted sensors that convert sounds into electrical signals directed to the auditory nerve – we will not discuss them here. Other solutions include systems that alert the user to specific environmental events through light, vibration or combination. The third group are communication technologies, which seek to facilitate a flow of communication and translation through speech, text and sign language in both face-to-face and remote contexts.

Using technology to enhance hearing

There are increasing numbers of features built into modern consumer devices that can be helpful to people with hearing loss.

Apple Live Listen makes an iPhone or iPad, act as a microphone and sends sound directly to AirPods. Live Listen can help to hear a conversation in a noisy area or where someone is speaking across a room. Alternatively, **Apple Sound Recognition** will notify you whenever your phone detects common noises that you might want to know about. **Apple Sensory Alerts** can also let you know of events in your preferred way. You can choose visual or vibrating alerts for incoming phone and FaceTime calls, text messages, mail, and calendar events. You could set an LED light flash for incoming calls or have your iPhone display a photo of the caller. You can also ask your Apple Watch to give a gentle shake whenever a notification arrives if you are travelling.

Google combine some of these features into Live Transcribe and Notification. This makes everyday conversations and surrounding sounds more accessible for people who are Deaf and hard of hearing, using an Android phone. Live Transcribe and Sound Notifications provides a free, real-time transcription of conversations and can also send a notification based on the sounds at your home. The notifications will make you aware of important situations at home, such as a fire alarm or doorbell ringing, to respond quickly. Even **Amazon Echo** can use sound detection to help. The smart speaker provides visual clues to those with hearing loss. For example, suppose it detects running water for some time. In that case, it can trigger the lights to flash on and off as a visual reminder to shut off the tap. Ultimately, Alexa's sound detection feature can help you and your home operate more smoothly. In addition, you can enhance your phones with extra apps that might be helpful if Deaf or if you have a hearing loss.

hearing loss or Deaf

The **SoundAlert** app turns any smart device into a high-tech alerting device. It can use the device microphone to pick up the environment's sounds and convert them into a visual/vibrating alert. These can take the form of onscreen notifications, vibrations and flashing lights. It can pick up a wide range of sounds. Some, such as smoke, carbon monoxide and gas alarms, are pre-installed on the app. Others can be added by the user according to their own appliances, e.g. doorbells and intercoms. The app does not need to be connected to the internet and can help those with a range of hearing impairments to feel safe and secure.

Sound Amplifier is an app for Android phones that improves the sound quality from a device, using headphones to improve listening clarity. You can use can Sound Amplifier to filter, augment, and amplify sounds around you and on your device. It can increase some important sounds, such as conversations, without over-boosting distracting noises. With two sliders, you can customise sound enhancement and reduce background noise.

Streamer is an app that allows you to create and receive a private and secure website, specifically for you. You can use your website as often as you like to caption and translate conversations with as many people as wanted. Streamer has features to support communication needs in many situations.

Subtitle Viewer is an Android app designed to find, download and display subtitles on your s smartphone. You can search for subtitles linked to a specific film or TV show. Once downloaded, the subtitles are synchronised to what the user is watching and playing in real-time. The subtitles slowly scroll up the screen, highlighting the current phrase or sentence. Because Subtitle Viewer works on a smartphone, it is fully portable. So you can view the subtitles wherever you view a film, whether at home, in the cinema or even on a train. Furthermore, many hardware devices directly enhance hearing or add extra functions to hearing aids. These would include:

Induction loops and digital hearing aids

Amplified TV headsets are intended for people who don't need hearing aids but perhaps have a mild hearing loss that makes it difficult to hear the television. They consist of a wireless headset that connects to the TV, usually through its headphone socket. The user can hear the TV through the headset and adjust the volume and tone of the sounds they hear. An **Amplified TV Headset** is very useful when more than one person is watching TV. The user can hear the sound without the volume of the TV having to be increased for the other people in the room. Several headsets can be linked to one transmitter so multiple users can watch and hear the same programme, but each can adjust the sound to suit their personal needs and preferences.

If someone has a hearing impairment, they may find a **wireless sound streaming system** can be useful. These normally include a transmitter with a built-in microphone and a receiver. Wireless systems can be used when having a face-to-face chat at home or in a pub or café by pointing the transmitter towards other speakers. Wireless sound streaming systems can also help when listening to music or television. This means that users can enjoy music with others without increasing the volume for everybody. There are many different brands of wireless sound systems, such as Oticon, Phonak and Bellman. Some come with listening aids included as part of a package. Some are designed to be paired with your own aids. Hearing aids in such settings can be made more productive by using an induction loop. An example is the **Bellman Maxi Pro Personal Amplifier** which can help you hear sounds and voices more clearly in group settings, with family, while watching TV. It also helps have clearer conversations and may improve everyday communication. The Maxi Pro connects directly to your TV with a streamer unit.

Hearing loops are a technology that transmits an audio signal wirelessly to a loop receiver device or suitable hearing aid. There are broadly three types of loops: fixed that are built into a room, portable that can be moved from room to room as needed, and personal which follow the hard of hearing person. One example of a personal system is the **RogerPen.** When spoken into, this acts as a simple microphone. However, when placed on its side, it becomes an omnidirectional microphone listening to all the sounds in the environment. One extra feature is that when placed into a docking station, it can connect to an audio feed, such as a television or even a PA system and transmit directly to the user.

FM transmitters

Similar to induction loops, FM transmitters can be valuable in helping facilitate a shared experience. Especially in social settings such as watching television, streaming video or on a large screen. These use a different method of transmitting signals and operating using FM similarly to radios. The **FmGenie Radio Aid** System is one such system that integrates with a wider range of technologies. In addition to these, other applications and software offer different ways to help mitigate the impact of hearing loss. For instance, some people find real-time text communication to be helpful. These can be dedicated hardware devices such as the **UbiDuo** or may be software apps that connect two or more mobile devices to facilitate communication. Messaging products may be effective in allowing individuals or small groups to communicate through text. Mobile and portable technologies allow partners with no hearing loss to speak to their devices. This is transcribed into text and shared with the group. Similarly, the text responses of the deaf person can be spoken out loud and amplified to the wider group. Speech recognition such as Google Translate can be useful in less formal settings where there has been little opportunity to prepare for the conversation. Google Translate can interpret speech to facilitate communication, and new versions feature a conversation mode where two-way communications can be supported.

Alerts, alarms and notifications

In many buildings, alarms and alerts are made by loud audible alarms. Fires and other emergencies are all delivered in this way. For people who are deaf, this may place them at some risk, and so a range of visual and tactile alert systems have been developed. Often these take the form of dedicated sensors and alarms that include a strobe light or vibrating units that can be placed under a cushion or pillow to alert the deaf person. More recently, these have been further developed as a wearable that the person can wear on the wrist to alert them to sounds in the environment around them.

The **Agrippa Deaf Alert Pillow Fire Alarm** and **FireAngel Pro Connected Alarms** are good examples of such alerting systems for emergencies in the home. In addition, devices including the **Amplicomms TCL350 Extra Loud Vibrating Clock, Geemarc Wake 'n' Shake Star Alarm Clock**, or **VC10 vibrating analogue alarm clock** are simple but very effective ways of ensuring that you are ready for an appointment. More recently, systems such as the **Apple Watch** have included tactile and vibration alerts. These, combined with a text alert, can warn of sounds in a room or notify you that messages are waiting on your phone.

More complex alerting and security systems are also available. For example, the **Jenile Security Alert System** consists of different sensors connecting to a centralised alert unit via Bluetooth. When a sensor is triggered, the alert unit displays a bright red light to give a visual warning to the user or can activate a vibrating unit placed under your pillow. In addition, there are various types of alert units. These include light cubes which use LED lights

hearing loss or Deaf

to indicate the source of the alert. The sensors include a Smoke and Fire Detector, a Gas Detector and a Flood Detector. The system also includes a Motion Detector and a Door/Window Detector, which will send a pre-set light or vibration alert to warn of intruders. Other systems such as the **Kidde 10SCO Combination Smoke and CO Alarm** or **Signolux Alerting System** offer similar features.

Having some devices designed and dedicated for a single purpose can be reassuring. For instance, many Deaf parents have found the **Digital Video Baby Monitor Watch** a great asset. The watch vibrates when a sound is heard in the baby's room, and a camera displays a video on the watch to see if there is anything you need to do. Many parents without hearing loss find this to be equally helpful.

Systems that can alert you of visitors can be useful as well. Some such as the **Bellman Visit** are designed especially for people who are deaf. Some others, such as the Ring video doorbell, have helpful features for some people.

Remote captioning and remote interpretation

For many who are Deaf or have hearing loss, remote captions and sign language interpretation can be a useful additional tool. Whilst some subscription services offer on-demand support, many such services require the booking of support in advance. As a result, they are most used in more formal settings such as planned meetings or group events where support can be made available. Apps for your phones such as **AVA** offer a choice between automatic captions or human captions, depending on how accurate you need them to be.

Human interpretation of speech into text and vice versa is also available using **Relay UK**, which is available through an app or a textphone. Connecting to an interpreter, they will provide you with text of what someone says or read out your written messages to the hearing person. Similar services such as **SignVideo** are available for Sign Language users. These are sometimes referred to as Text Relay Services (TRS) or Video Relay Services (VRS).

British Sign Language support

For those with little or no hearing or are Deaf, there are also technologies and applications based upon support for British Sign Language and other sign languages. While some applications seek to translate text to BSL or BSL into speech, these should be treated with a degree of caution and fully considered before relying upon them. However, some additional technologies and services use connectivity to aid BSL communication remotely. For example, video conferencing tools, including Skype, WhatsApp or Zoom, have allowed Deaf people to communicate with one another using BSL. Equally, they can facilitate communication between Deaf and hearing people by including a third person as an interpreter. Some services are available that make this easier to achieve. For example, **SignTime** is a new service from Apple that allows Deaf people to directly connect to Apple interpreters for BSL and other sign languages from their phone or tablet. The service aims to be available to other suppliers of goods and services in the future.

*

There are many options available to support people who are Deaf or have hearing loss. However, it is important to carefully consider the extent to which you can hear sound with or without hearing aids, the challenges you want to address, what technology you already have, and where the challenge occurs. Having thought about these issues, the technologies listed in this toolkit offer a starting point to making decisions.

neurodiversity/cognitive disabilities

For people with neurodiversity and cognitive disabilities

The area of cognitive disability is complex and is increasingly viewed within a framework of neurodiversity which includes people diagnosed or who identify as dyslexic or autistic. It is useful to suggest a distinction between those who have difficulty processing written text and those who experience other forms of cognitive need that have an impact.

Reading and writing

Reading and writing needs are often characterised as including those people with Dyslexia. There is no sensory or physical barrier to reading and writing in these cases. Still, those with these needs experience a barrier to processing text creating challenges to understanding and creating written documents. In such cases, technologies that support or review text that has been written and present or enhance text to support ease of understanding are especially valuable.

Some people who are neurodiverse find both reading and writing of traditional text to be challenging. This chapter explores technologies that enhance text to make it easier to comprehend and produce. These include features built into a device, mainstream products that address limited literacy, and products designed to help those with reading and writing or print impairments. It is not unusual for people to use a combination of these.

Magnification, fonts and colours

Making the text more readable can start with some simple adaptations. Adjusting the size of the font, the contrast, the colours, and the choice of the font itself can have a significant impact. There is no single combination that is ideal for everyone. The most helpful varies from person to person and even the time of day, including how long anyone has been viewing a screen. A scanner and OCR software such as **OmniPage** can scan a document and then change its appearance digitally for printed pages. Details of how to make these changes can be found in the **Ease of Access Centre** in Windows, and Microsoft Word allows you to change the background and fonts to make them easier to read. Good advice is to present documents in 12pt Calibri font, with a high contract text on the background. For those struggling to read the text, magnification built into most devices can help reduce clutter and make words easier to follow. These may be referred to as **Magnification** or **Zoom** features.

Reading aloud

Many people find that they need further support to help with reading text. One of the most useful tools is to have the text read aloud. This technology is increasingly built into our devices for when we cannot safely read from a screen, such as when driving. In Windows, this is available in two ways, either using the built-in text to speech feature **Narrator** or the Read Aloud option in Microsoft Office tools. Read Aloud is usually found under Review in the application toolbar. Similar features can be found in Google docs and Adobe Acrobat.

Going further, we can use Microsoft's **Immersive Reader**, a free tool that offers several features to improve reading and writing for people, regardless of their age or ability. For

neurodiversity/cognitive disabilities

example, it lets you choose fonts and change size, spacing, and colour. It can also read text aloud.

Sometimes, however, it is helpful to take a document and create an audio version as an MP3 that the person can take with them to listen to at their convenience. For example, **Robobraille** is a useful website that allows you to upload a document that needs to be read and quickly receive back a version as an MP3 or other format. In addition, some people with literacy needs find it easier to listen to a short document than struggle to read it. Robobraille provides a cheap option.

Grammar and spelling

Many people with literacy needs correct spelling and grammar. They may be familiar with basic spellchecking tools in Office. Still, there are more advanced features and applications that can be useful. **Microsoft editor** is a feature in Office 365 with several features. These allow you to check a document for spelling, grammar, clarity, conciseness, formality, conventions and vocabulary. These tools can be found on the Home tab of the ribbon and help produce a high-quality document that accurately reflects your thinking. Extra tools such as the **Grammarly** app have more powerful versions of these features. These can be tailored to the target audience and includes a plagiarism checker. This may be helpful when producing reports and other documents to protect you from accusations of using someone else's work. Grammarly is especially helpful in proofreading documents that many people find difficult and stressful.

Word prediction

Producing accurate words and text is also assisted by using text or word prediction. This has been a specialised feature for people with severe writing impairments. Still, it has become mainstream through our phones where we write using a small keyboard, so reducing key presses helps us write more efficiently. While mainstream applications and operating systems can offer some forms of word prediction, there is still a strong case for more specialised products such as **ClaroRead** or **TextHelp Read and Write Gold** which combines tools for reading and writing into one interface.

Voice recognition

Typing can be slow and difficult for many people, especially where keyboards are small and mistakes easy to make. Speech or voice recognition and dictation applications can help address this. This is very easily accessible on phones, where it is increasingly used as a standard alternative to typing. For example, the keyboard has a small microphone icon to open a dictation window on the iPhone and iPad.

All major vendors have some form of voice recognition integrated into their operating systems and applications for computers. For example, in Microsoft Office, a Dictate button is usually next to the editor on the Home tab of the ribbon. In Google Docs, Voice Typing can be found on the dropdown list from Tools on the main toolbar. Voice recognition is powerful but may not be suitable for everyone. A headset or high-quality microphone is recommended for computers, and anyone using it is strongly advised to check their text carefully before completing and sharing a document.

Memory

Technology can be a valuable and convenient way of supporting anyone experiencing

neurodiversity/cognitive disabilities

problems with their memory. These include prompts, reminders, alerts, and notifications alongside technologies that allow us to record our activity to recall later, remember people and find mislaid objects. Choosing the best app and service for an individual will be based upon many factors. These might include ease of use, functionality, purpose and cost.

Prompts and reminders

Most smartphones have a built-in clock and calendar. These offer a starting point for anyone needing regular prompts or reminders of what they need to do and where they need to be. By sharing a calendar, others can put appointments into a calendar which then launches a prompt or reminder before it is due. There are also different smartphone reminder apps that remind us of what we need to do. Many integrate with the calendar and send notifications when a deadline approaches. Although each has different features, they share a common purpose to help us remember important things. Reminder apps help people create to-do lists and can include tools that help list tasks by priority, so we know which things to focus on first. Some good starting points are **Remember The Milk,** which helps manage tasks, works across devices and can be shared with others, and **Microsoft To-Do,** which creates lists for what should be remembered and includes a smart suggestion feature to learn habits and offer suggestions for things you might need to do in the future.

Alerts and notifications

These are technologies that offer alerts and notifications. They differ from prompts and reminders in that they are used more to send an alert or message based upon a specific trigger. These might be location, a deadline or some other important information. They can be as simple as an SMS or instant message to explain and remind us what to do. For instance, a delivery service may have had a message saying that the delivery cannot be accepted. Both inform their driver and remind them what to do as a result.

These are individual to the person with memory loss but might be deployed more widely in some cases. For instance, an app that tells someone that they are parking in a high crime area and to take extra care to set the vehicle alarm might be one option. They can also be useful for lone workers who are expected to check in to confirm they are safe and receive an alert, having missed a deadline. Some useful starting points are **Pushover,** which will send real-time notifications to Android, iPhone, iPad, desktops and wearables. Finally, **CallMy** can be used by an individual or an organisation to send urgent notifications and receive distress alerts.

Recognising people

One aspect of memory that is discomforting and creates anxiety in the workplace and daily life is an inability to remember the people you meet and speak to. It can be helpful to have a notetaking app on a phone where a person can make notes following an interaction. Still, an app that helps organise and retrieve those notes is even better. One of the better-known apps that tries to support this is **Revere**. Revere will help people remember the last conversation, their favourite drink or the names of family members. Revere can also set reminders to help people remember how and when to stay in touch.

Activity recording

For some people remembering what they were doing and when can be challenging. This may prompt other memories, such as 'where I left my coat', but can also clarify how you use time,

neurodiversity/cognitive disabilities

improve time management and reduce stress and anxiety. Activity and time tracking tools allow the person to record what they have been doing as they do it and then to review that as and when needed. Good starting points include **ATracker**, a time-tracking application that is easy to use and requires minimal setup. You see the complete task list on the main screen, with an overview of time and goal progress. You start or stop recording by tapping a task. Equally, HourStack can help plan, track and prioritise time through one visual calendar. It is easy to log time spent and review what has been done and when.

Finding my things

One of the most common causes of stress from a poor memory is when you cannot find the things you need when you need them – keys, wallet, coat and of course your phone itself. Technology will allow you to tag your possessions and then find them on a map or by using a tag to emit a sound to locate it. For example, both Google and Apple allow you to set up a group or use a webpage to make your phone ring or locate it on a map. Other objects can also be tagged using RFID tags attached to the object and which can then be located using a phone. One of the best-known systems is **Tile** which offers a range of tags of different sizes and formats that can be stuck to an object, placed inside an object or hooked onto an object.

Personal organisation

For some people who are neurodiverse or have cognitive disabilities, challenges may exist concerning personal organisation at work, home and education. Traditional approaches such as lists and reminders may not be sufficient. Many tools provide further support across different settings. In choosing the most suitable form of support, it will be important to think about the information that needs to be organised, the individual's specific needs and the setting where the support will need to be applied.

Information curation

It is not unusual for anyone to feel overwhelmed by the volume of information that is readily available. Organising or curating that information to store, search, find, retrieve, and use it is increasingly important at work and in learning. Fortunately, there are tools available to help. These include **Google Keep** and **Microsoft OneNote**. These are apps and software and also offer plugins to your browser. When you identify the information, you need now or later, you stag them on your device to store that information. You can think of them as digital notebooks that can be added to from any location. Once you have the information stored, you can search and use the information at any later point.

More features are included in **Evernote,** including more advanced search and retrieve functions. These allow you to search by title, by the contents of the text or by a tag that you created and associated with the information. The information you can store is not limited to web pages but includes videos, YouTube pages, PDFs and images. Evernote is a powerful and useful tool for those who need access to a lot of information but find it difficult to maintain that information.

Mind-mapping

Graphic thought organisers are a different way of organising information. These can include ways to record our thoughts and ideas. These may be referred to and include mind mapping. In these tools, we place ideas in a bubble on paper or onscreen rather than simply making lists. We then draw lines to connect each idea to others or add extra thoughts to the original

neurodiversity/cognitive disabilities

concept. Mindmaps can represent tasks, words, concepts or items related to a central subject. Mindmaps can be useful for Planning, Organising, Note Taking and Studying. Good starting points to introduce mind mapping are **Xmind** and **Freemind** for the PC or, for more advanced features, there are **Ayoa** or **Inspiration**. A simple online tool is **www.bubbl.us** for both individuals or small groups. Tools such as **Claro Ideas** are especially helpful in allowing ideas to be represented both by text and images.

Scheduling assistants

As well as organising information, we also must organise our time. This is more than simply remembering an appointment or a time for an event. It is about organising the steps to be completed to arrive on time. One of the most popular is **Todoist**. This provides an overview of everything we need to do and tracks the tasks. You can sequence the steps that need to be taken and schedule steps from one or more tasks into a day.

Visual schedules

Todoist requires a high level of confidence and competence in the user. Other, simpler tools to help with scheduling include visual scheduling. Visual scheduling supports those with autism or other cognitive needs through a graphic representation of tasks and activities. They are useful for breaking down tasks with multiple steps and ensuring that rules and deadlines are understood. In addition, they can help reduce anxiety by providing consistency and photographs, icons, and symbols can all be used to build a schedule. Technology can allow such schedules to be constructed digitally. This lets you carry the schedule with you and make any changes as needed.

One useful app available for iPhone and iPad is **Visual Schedule Planner.** Visual Schedule Planner offers an Audio-Visual Schedule or Calendar for those who benefit from visual supports. By using different formats and media, the prompts and events can be represented in the style preferred by the individual. Alternatively, **HandiCalendar** is a smartphone app to support you in planning and navigating daily living. HandiCalendar provides users with a visual and audible overview of their day, week and month. Activities can be assigned images, and alarms can be set to signal their start and end times. A countdown feature uses a row of dots to indicate time elapsing. In addition, there is the option to have text read aloud by a speech synthesiser.

Integrated digital support systems

Suppose a range of technologies to support personal organisation and self-management are required. In that case, it may be appropriate to consider **Brain in Hand.** This is an integrated approach to address the diversity of challenges that an individual faces. Brain in Hand can be personalised, is always available through a smartphone, tablet or website, connects to additional support, and can help users to manage and organise the things that are difficult whilst working towards goals

For many people who are neurodiverse or who have a cognitive need, there is a need for tools that directly address the challenge they face and tools that provide a scaffold around them. These scaffolds provide support and advice when needed. In addition, there are significant overlaps with technologies associated with other needs. These include communication and mental health needs as they may result from more than one form of impairment, and people with different conditions may share similar challenges.

SOUTH LONDON

WE DEVELOP DIGITAL SKILLS & CONFIDENCE IN THE COMMUNITY

ClearCommunityWeb help people feel more confident and comfortable with technology through classes, workshops and individual support.

Community Learning
Workshops
Drop-in Support
Home Visits

Contact us to see how we can help:
info@clearcommunityweb.co.uk
07523 646 277

clearcommunityweb

For people with communication impairments

Assistive and accessible technologies support communication in a variety of ways. For many people, technology is at the heart of their communication ability, reducing social exclusion and allowing them to participate in their community. In addition, such technologies can support both formal and informal communication with peers.

Communication methods

Everyone uses augmentative communication every day. For example, we clarify words through facial expressions, gestures, and body language when holding a conversation. These can add to the meaning or reverse it completely. For example, a yawn may completely negate what you are saying. For people with disabilities involving speech, many methods support and augment communication. These include individual methods of sign and gesture, standardised signing and symbol systems and electronic devices. Developing and using an AAC (alternative and augmentative communication) system may be a lengthy process. Training and practice will be needed, and materials prepared and kept up to date. These processes involve everybody – any communication system must be mutually understood.

Systems and methods

These terms are sometimes used to mean the same thing, but there are some distinctions. For example, the 'English language' is a communication *system*, but speech and writing are two *methods* to use that system.

Symbols, text and speech devices

Some people can have difficulty with face-to-face communication for many different reasons. For example, physical disabilities and motor coordination problems can make speech production difficult or impossible. In addition, people with some types of learning difficulties can find it hard to produce speech or handle spoken language. The term AAC (alternative and augmentative communication) is used to describe the different methods that can be used to help people with disabilities communicate with others. As the term suggests, these methods can be used as an alternative to speech or supplement it. No matter what their difficulties are, few people can be said to have no method of communication. However, many people will have difficulty getting their message across. This requires effort from the listener or communication partner. Communication is essentially a two-way process that must involve some degree of mutual understanding and a commonly agreed method. Even when two people can easily talk and understand the same language, misunderstandings and failed communication can occur.

Symbol systems

A variety of symbol systems are available. They have been developed for people who have difficulty understanding written language, such as those with learning difficulties or younger children. Systems can be combined with personal symbols, objects and photographs if this helps.

Symbols can be useful for longer messages and are quick and easy to recognise. We can see this in our daily life and the world around us. Symbols can be presented in many ways, including technology devices, paper charts, and communication books. They can be produced by drawing, photocopying or using an app or software to print out charts. Examples of symbol systems include Arasaac, Mulberry, Makaton, PCS and Bliss.

Signing and gesture systems

Manual signing has been in use for many decades, including for people who are Deaf. However, different systems have been developed to meet the needs of people with cognitive or physical disabilities. Signing systems have the advantage of not requiring any additional equipment or materials but can be harder to learn. Nevertheless, there are many sources of information and training to help learn basic signing. Some programs even have signed support in the form of short videos.

Text-based

Written or computer-generated text can be used for messages. It can be produced on-demand or as pre-stored messages. Some examples of such products to assist with communication are suggested below.

Face-to-face communication

Technology can generate speech from a variety of inputs. Speech output is available in various languages and voices, and many people take time to choose a voice that they feel best represents their personality. The speech that a machine generates may take time to be produced as a person with a disability will have to type or create a sentence and then ask the machine to speak it out. Some people prefer to have each word spoken individually, but this can be more difficult for the listener to follow as each word may take seconds to produce.

Regardless of the personal preference of the person with a disability, it is important to remember that machine-generated speech will be slower than natural speech, and the listener or communication partners will have to be patient

Using text to communicate

There are many ways a person with a disability can produce speech by typing. These broadly fall into two categories. Specialist devices and apps for use on smartphones and tablets. Both are widely used. Many devices and apps are suitable for people to use. But it is worth examining two examples to illustrate the difference.

A **Lightwriter** is an example of a dedicated device developed solely for spoken communication by typing on a keyboard. ClaroCom is an example of an app that can be downloaded onto a phone or tablet and supports speech output by typing on an onscreen keyboard. Such applications are often free or very low cost. Moreover, the user can easily carry them on their phone without carrying another device. Increasingly, those who have lost their voices are turning to apps on mobile devices as they are more familiar with the interface and use of the device.

communication impairments

Using images to communicate

In recent years we have seen a huge growth in using images in online communication. Emojis' increased use to communicate complex thoughts and clarify or add tone to a text is well documented. Many people now enhance their communications through images from a simple smiley face to emojis to denote purpose and even suggest a disability. However, for some people, graphics and symbols are at the forefront of their preferred communication to increase communication speed or where reading text is challenging. Symbol sets differ from location to location, and new symbols sets are being developed to support different languages and cultures on an ongoing basis. Some of the best known and widely used symbol sets include Bliss, PCS and Rebus.

Graphic symbol sets

Different symbol sets have different rules guiding the design of a symbol. Some are abstract, whilst others are based on a graphic representation of objects or concepts. For instance, the word 'love' is represented differently depending on the set selected. Symbols work most effectively when they feel familiar to those involved in communication and interaction, so in some cases, different symbol sets have rules that reflect the culture of the users. One such example is Tawasol, a symbol set developed in the Gulf region for Arabic speakers. (www.globalsymbols.com).

Symbols are an important communication component through technology for those without a voice. Symbols can be selected from designed grids, where each symbol is linked to a text and spoken word equivalent. A user can hold a conversation using sentences by stringing together such symbols. Symbols and graphics can be used in two distinct ways, but both should be available. The first is to use technology to create printable boards that can be laminated or put into a book to communicate at any time. Where a person has a physical disability, these need to be carefully positioned to allow them to point at them with a finger, hand or with their eyes. Low tech resources such as the **PicSeePal** and or an **ETran Frame** can help with this.

Global symbols is a website that you can use to design printable boards for communication. All symbols and technology are free to use and share with other people. This is a very simple and low-cost way of getting started with symbol communication. The boards created can also be saved and stored for use with some communications aids and apps.

Communication aids and apps

Communication aids are sometimes referred to as 'voice output communication aids'. They can be dedicated devices designed only to work for communication or apps installed on a tablet or phone to provide similar features. In most cases, the apps and devices can be accessed in various ways depending on the person's physical ability.

Communication aids

These range from simple one message devices such as the **Big Mack,** which looks like a large switch but can be used to record simple messages spoken aloud when the button is pressed. There are also aids with multiple cells, 6, 9 or 12, such as the **GoTalk**. With these devices, different messages are stored for each cell. The message is read aloud when the specific cell is activated. More complex communication aids include the **ProxTalker** or **eye-tracking aids**. These combine cells on a screen that speak out messages, with cells that link to different collections of symbols described as boards or grids. These different grids or boards can have been designed for specific settings or topics.

communication impairments

Communication apps

These dedicated devices were very reliable, but they were also very expensive since they only had one specific use. As a result, many AAC (alternative and augmentative communication) apps and software have been developed to run on laptops, tablets or phones in recent years. Some of these also have versions that can be used with a wearable device or eye-tracking to select cells. Choosing an app or aid is complex, but some useful starting points include free solutions such as the **boardbuilder** from Global Symbols, **cBoard**, **OTTAA, Coughdrop**. There are also commercial products with powerful features available, including **proloquo2go, Grid 3** and **Avaz** amongst many others. However, before choosing an option, it is recommended that professional advice is taken to compare features and guide a decision. Sources of such advice are included later in this toolkit.

Communicating and interacting online

For many people, the internet provides a powerful means by which they communicate and interact with peers. Whether this is to arrange face-to-face meetings and gatherings or simply as an end, online communication is an important part of social life for many.

Instant messaging and VoIP (voice over internet protocol) solutions offer multiple ways to communicate, making them ideal for communication over a distance.

Using messaging and VoIP

Skype is an example of an instant messaging solution that offers multiple means of expressing yourself. Skype offers text chat, voice and video chat. In addition, it allows images and emojis to be added to conversations to add clarity or tone to a conversation.

Skype allows group conversations, so it can also be used for people with little or no hearing by adding a sign language interpreter to the conversation. They sign the spoken conversation and can speak for a Deaf person signing. This multimedia approach to communication greatly benefits users and is relatively easy to implement.

WhatsApp

WhatsApp is a free messenger app for smartphones. WhatsApp uses an internet connection to send messages, images, audio or video. WhatsApp uses an internet connection to send messages the cost is low. It is extremely popular in many parts of the world where free wifi and low-cost data packages are available. It now has one billion users worldwide and is the biggest online messenger service. It is especially popular with young people with features like group chatting, voice messages and location sharing.

Social media

Many forms of social media are used by people with and without a disability, contributing to a feeling of belonging and inclusion. Some of the most widely used include the following.

Tumblr allows users to share text, photos, quotes, links, music and videos from a browser, phone, desktop or email. It has been described as a cross between a social networking site such as Facebook and Twitter and a blog. Some people refer to it as a 'microblog' as people post short submissions instead of longer reflective or diary-style pieces found in more traditional blogs.

Reddit is essentially a huge forum categorised by various topics called subreddits. These are usually preceded by 'r/' – for example, some of the straightforward subreddits include r/news or r/books. Others are a little more unusual. If you have an interest, there is probably already a subreddit about it. Subreddits are run by moderators, who may have their own rules for participation in the subreddit.

Facebook (also known as Meta) is a social network where users post comments, photographs and links to news or other interesting content on the Web. It also has the functionality to allow users to play games, chat, and livestream video. Content can be made publicly accessible or shared only among a select group of friends or family or with a single person. Facebook allows users to maintain a friends list and set their own privacy settings to determine who can see the content. It allows users to upload photos into albums that can be shared or made public. In addition, it supports online chat and the ability to comment on 'friend's' profile pages to keep in touch or share information. Facebook supports group pages, fan pages and business pages. In addition, businesses use Facebook as a vehicle for social media marketing.

Twitter is an online news and social networking site where people communicate tweets like Tumblr in short messages. It is sometimes referred to as micro-blogging. Some users use Twitter to discover people and companies of interest online and follow their tweets and thoughts.

Twitter is easy to use, and you can review the content at a glance. Tweets are restricted in size, ensuring tweets are easy to scan. In addition, every 'tweet' is limited to 280 characters or less. Which encourages simple and clear use of language. You then choose to read your daily Twitter feeds through various Twitter readers, including **EasyChirp**, optimised for those with little or no vision.

Instagram is a social network for sharing photos and videos from a smartphone. It is similar to Facebook or Twitter, where users create an Instagram account with a profile and news feed. Every photo posted is displayed on that profile. Users who follow that account see the posts in their feed. Like most social networks, users interact with each other by following them and by being followed and then commenting, liking, tagging and private messaging. In addition, photos seen on Instagram can be saved onto a device.

Snapchat is a messaging application for sharing photos, videos, text and drawings. It has become hugely popular in a short space of time, especially with young people. Like Instagram, it offers a range of effects and filters to enhance pictures which is appealing. One feature that makes Snapchat different from other messaging apps is that the messages disappear from the recipient's phone after a few seconds.

Pinterest is an online pinboard where users can collect multimedia such as images. Users can create themed boards for their pins, and each board can reflect an interest. Others can follow the user or a specific board and then interact through liking, commenting and repinning posts. Pinterest has very strict community guidelines on what can be posted. As a result, it is one of the safer social networks.

*

Communication is a basic human right. For those without a voice, great care should be taken to ensure that they have the support to speak out, advocate, and take part in everyday life. This toolkit section offers some initial starting points and many issues to consider. If you wish to investigate this further, we highly recommend undertaking the free training at training.globalsymbols.com, or exploring some of the additional sources of support included later in this toolkit.

For people with mental health needs

Since 2019 more people than ever have reported experiencing mental health issues. These can be debilitating and impact entire lives, limiting progression and achievement at work and in education and severely restricting the quality of life and social interactions.

In an era where ideas of the workplace are evolving, mental health can be a major cause of lost productivity, whilst students in schools and colleges are reporting mental health needs at an alarming rate. The pandemic alongside structural change, systems under stress, underemployment and failing institutions have had a tremendous human cost in increasing anxiety, anger and loneliness.

Both individuals and organisations are turning to technology to proactively support people with mental health needs to reduce the likelihood of a crisis. Technologies with value include productivity applications that help reduce anxiety in presenting ideas and functionality that recognises events and interactions that trigger a crisis and supporting self-management by recognising warning signs, identifying actions that prevent a crisis from developing, suggesting coping strategies during a crisis and tapping into sources of support including peer groups of people experiencing and managing distress.

For many people, daily life can be characterised by periods of stress and anxiety. For people with a disability, stress and anxiety can rapidly become a second form of disability that they experience. The tools and technologies in this section offer an approach to self-management by recognising triggers and suggesting ways to mitigate anxiety or panic. Self-management is a vital part of creating a support scaffold around a person with mental health support needs. Still, it should be seen as the first stage of support and routes to additional help should be considered.

Focus and avoiding mistakes

The fear of making mistakes or simply submitting poor work that requires a lot of editing and proofreading can impact confidence in the workplace and add to anxiety for many people with disabilities. Some simple scaffolds to support the individual can be useful. It is helpful to refer to the materials on technology support for reading and writing in earlier chapters. Some of the most useful that are discussed include the **Editor** function within Microsoft Office and software such as **Grammarly,** which checks and proofreads documents as they are written.

In addition, some people find that potential distraction online, in webpages and social media to be a serious issue. One way to help reduce this and retain focus can be found in the **Ease of Access Centre** for Windows as Focus Assist**.** Focus Assist helps by reducing the number of notifications sent to the desktop and hence help avoid distractions from tasks. Other useful tools include features under Display, which include turning off animations in windows. Others find that adding an extension to Chrome such as **Blocksite** can help avoid the temptation of online distractions.

Recognising triggers and self-management

A panic attack or overwhelming anxiety can often creep up on a person. It can be difficult for an individual to recognise warnings and indicators that they need to do something different to avoid a repeat of a crisis that they had experienced previously.

Apps such as **Worry Watch** or **Serenita** can offer a useful means by which we can record our activities, including interactions with other people or workload and our moods. We can identify patterns and try to self-manage the stress earlier in the cycle by reviewing these. Self-management is a well-regarded approach to helping those who experience such anxiety.

For example, Worry Watch suggests that we record by writing down what has bothered us and set a reminder to revisit it later. We can reflect upon it after it is over and consider whether the outcome was as bad as feared. We can use the dashboard to reason and consider our patterns of behaviour and thought and then refute our anxiety by realising that many worries were often unfounded. Similarly, Serenita works by helping to monitor stress levels and then offering personalised exercise to achieve stress reduction impact and real-time feedback to help ensure that the exercises are used correctly. **CBT Thought Diary** is a free app for Android. The app can help you keep a diary of your anxiety and depression and your health, including your heart rate, number of steps, and sleep duration.

Mindfulness

Having identified that we are at the beginning or even quite advanced in a cycle of anxiety, the apps will often recommend some actions. Taking a break, stepping away from a situation or changing something we are doing. However, as anxiety builds, it may be important to also have a strategy to increase the feeling of calm and reduce tension. Apps such as **Breathe** and **Smiling Mind** offer simple mindfulness activities that help clear our minds of anxious thoughts and encourage a slower heart rate and breathing. Smiling Mind is a good example of such technology. It offers many exercises tailored to different demographics and desires. Moreover, all the content is free, making it ideal for many beginners to see if mindfulness is helpful for them.

Other apps draw upon some elements of Mindfulness and combine them with other approaches and features. **SAM anxiety app** is one such example. SAM offers tools to help with anxiety when you need it. The features include breathing, mindful observation of pictures, and redirecting focus, which are valuable techniques. Most of the techniques are simple and clear, and interactive.

SAM also includes longer-term tools, such as increasing awareness of unhelpful thinking styles, questioning your thoughts, and self-care. The app includes a brief four-part scale to rate your anxiety, worrying thoughts, unpleasant physical sensations, and avoidance. The anxiety tracker offers a visual summary of your anxiety over time, helping you monitor yourself, a valuable element of change. One app that focuses on how anxiety presents itself is Headspace. Whilst offering an introduction to mindfulness, it looks carefully at your sleep patterns and offers suggestions to address any lack of rest.

Peer Support

Equally, however, there are times when some help and support from others are vital in

mental health needs

breaking the cycle of anxiety. Having a buddy in the workplace to go to when you are anxious or your app is warning you to step back is ideal, but such a person may not always be easy to find. Apps such as **Be a Looper** and **Wisdo** create small mutual support networks. Here we can record our feelings of tension and monitor the feelings of our peers. As a group, we can then reach out to each other to help when someone is having a difficult time. Be A Looper is a daily mental health check-in and peer support app to keep users 'in the loop' with five people globally. The tool helps users share how their day is tracking in a gamified way while also keeping a close eye on those they care about. Wisdo aims to provide a meaningful alternative to mainstream social networks by connecting people to those who share their experiences and are ready to offer non-judgemental support and encouragement.

Access to professional support

But there are also times when peer support needs more structure and experience. Solutions such as **Brain in Hand** are a valuable way of escalating a need for assistance when it is most required. Brain In Hand is not a simple app to use, and training and support may be required to help introduce all the functions. Choosing the ones that will be of greatest help will be important in working through the options together. Brain in Hand is a 'digital self-management support system' for people who need help remembering things, making decisions, planning or managing anxiety. It is often used by people who are autistic, those with learning difficulties or who are managing their mental health.

Technology has been suggested as a cause of the increasing rates of mental health needs in communities. However, increasingly, technology is an important and very familiar way of connecting, sharing, and accessing goods and services for many people, especially those under the age of fifty. The apps mentioned here are all available online. There are many other examples of software and apps for phones, tablets and computers that can be found through the Further Sources of Information links listed on page 234 of this toolkit.

In the workplace

Both accessible and assistive technologies are vital in the world of work. The technologies that support access to information, reading, writing and communication can all be applied to the world of work. However, as well as assistive technologies, you will need access to the business and productivity tools to complete tasks. For this reason, any assistive technology provided at work must be checked to ensure that it works with the business tools. Some business tools that you may need to use include:

Microsoft Office

The most recent version of Microsoft Office is Office 365. It consists of a range of applications for email, word processing, spreadsheets, and presentations. Each has many features to make them more accessible and usable by people with a disability.

Microsoft 365 apps integrate with assistive technologies and accessibility settings on many devices. Some applications have easy-to-access features or learning tools to help with literacy. These include accessibility checkers, templates, autogenerated Alt-Text for images, and captions for speech to help make emails, documents, presentations, and meetings more accessible. Some of the more useful features of the Office programs include:

Read Aloud reads all the text in a document using built-in voices. Many people find this useful when reading is difficult, including temporary needs such as eye strain.

Editor offers access to Microsoft's spelling and grammar tools as well as helping to check for clarity. As with any automated checker, you should review each suggestion as their advice is not always accurate

Immersive Reader offers a range of features to make reading easier. Changes to layout, speech output and a reading ruler can all be helpful for people who are struggling with large documents

Focus is a useful tool for anyone who finds it difficult to concentrate on work for longer periods. Focus removes all the distractions on the computer and allows you just to concentrate on your document.

Speech recognition is built into Office 365 (and other productivity tools such as Google Docs). It allows you to dictate everything you want to write. This can be an alternative to writing using a keyboard all the time. Still, others find it useful when their hands become uncomfortable. It can also be a useful way of reducing the risk of RSI and strain in your hands and arms.

Accessibility Checker allows you and your colleagues to check that documents and presentations will work effectively with assistive technologies. This is good for you and will help them produce accessible resources for customers.

Video Conferencing

Many video conferencing platforms are available, but **Zoom** is the most popular amongst many people with a disability. This is partly because it is one of the simplest to use without being unduly complex to log into a meeting. It also has quite advanced transcription features for those who struggle to follow the spoken conversation.

TBG
T Brown Group
A trusted name in property maintenance for over 50 years

T Brown Group is experienced in delivering servicing, repairs, maintenance and installations covering all types of heating systems including: Gas, Solid Fuel, LPG and renewable, including ground source, air source, solar thermal and solar PV

T Brown Group remains a family-owned business and is committed to its mission statement to be an industry leading service and maintenance provider, delivering excellence through creating innovative, efficient, customer centred solutions that deliver demonstrable best value to our clients.

Maintenance and repairs

The company contracts span a varied selection of property tenures including; social housing, sheltered housing schemes, assisted living schemes and supported housing. The scope of work undertaken includes gas installations, servicing, planned maintenance, responsive repairs and maintenance, and void works.

Committed to Social Value

T Brown Group places significant importance on delivering Social Value Commitments which are aligned to the immediate community agendas of the boroughs and Housing Associations they serve.

 @tbrowngroup   @t-brown-group

workplace

Technology is increasingly broadening the opportunities for people with a disability to enter the workforce. Whether through personal assistive technologies or through the growth of connectivity and online collaboration, much greater flexibility in employment is available,

Collaboration Software

Microsoft Teams is a messaging and collaboration application with features to help everyone contribute to conversations and work together. You can access meetings and participate from anywhere with Microsoft Teams desktop and mobile apps and Microsoft Teams on the web. This includes some core features such as live closed captioning (US English only), adding human-generated captions, gaining attention by raising a virtual hand, reducing background noise in meetings and adding an interpreter to a call

Teams supports you to communicate with co-workers regardless of language, cognitive needs or visual capabilities. You can limit distractions with Do Not Disturb mode, customise reading and viewing for visual and cognitive needs and send audio messages on Microsoft Teams mobile and use Windows dictation on Microsoft Teams desktop. Teams facilitates collaboration by adapting to learning pace, motor skills, and communication preferences. You can have documents read aloud and broken down by syllables with Immersive Reader and ensure content is easy for everyone to read and edit. Any content you need can be tagged and pinned to make it easy to find when you need it again.

In **SharePoint** in Microsoft 365 experiences such as accessibility features are built-in, including the core features such as document libraries, lists and pages. In addition, the features include keyboard navigation, support for screen readers, and extra keytips. For example, you can also open a list of keyboard shortcuts by pressing '?' (question mark).

SharePoint supports the accessibility features of most web browsers to help you to access and manage SharePoint sites. The browsers provide support for keyboard interactions, and those people who don't use a mouse can then use a keyboard to use the site and perform actions.

SharePoint provides a 'More Accessible Mode'. To activate this mode, open a SharePoint webpage and press the TAB key until you find the 'Turn on More Accessible Mode' link. This feature recreates the webpage in standard HTML, making it friendlier for screen readers. For example, in More Accessible Mode, drop-down menus are converted to hyperlink lists and objects to allow screen readers to understand the content.

Dropbox offers cloud storage and file synchronisation so you can access your files from anywhere. Some of the most important features include file storage, on-demand access to files from any device, and Secure sharing. It also allows businesses to create Team Folders where content is owned by a group rather than an individual. The content automatically syncs for all group members. Each version of Dropbox has similar accessibility features and works similarly with your assistive technologies. It meets most of the accessibility standards. While it may take a little time to become familiar with it, it is a useful and accessible approach to collaborating with colleagues.

Accounts software

Software to help manage financial tasks from invoicing to accounts, managing expenses and making payments is vital to most businesses. However, it is an area where many packages struggle to be fully accessible for some people with a disability, most notably those with little or no sight. Most packages work reasonably for people with other needs. They may have keyboard shortcuts and work well with different mouse and keyboard alternatives, including

dictation to enter details, but not always to manage menu items.

Here are some impressions of some popular accounts packages for you to think about:

Xero is a web-based package that has become extremely popular in recent years. However, whilst it may work well for some disabilities, Xero is not fully accessible for people with limited vision.

QuickBooks is not fully accessible out of the box. Although it does have some features such as contrast and colour modes and the ability to resize text, that are helpful. QuickBooks can be made to work with both the screenreaders JAWS and NVDA, but you will need to download extra scripts to use these. This can be complex, and some professional support is recommended.

Sage 50 accounts software has been in use for over 20 years. Although Sage has committed to making more products accessible through a policy and training, there is little suggestion that this process is complete.

FreshBooks has been reviewed by blind users who report that they find it possible to track time and expenses related to projects at any given time. FreshBooks makes sending detailed invoices easy and can be integrated with PayPal or Stripe to provide various payment options for the customer. The most accessible version appears to be Freshbooks Classic which has been shown to work quite well with screenreaders, including those built into iOS and Android devices.

Travel and Transport

Whilst many jobs may allow you to work from home for at least part of the week, many need you to be in a workplace to do your job. So, travel planning is an important part of getting to work. The best way to get to the workplace is by taxi for some people, and Uber and similar apps have proven popular with many. A regular destination can be saved as a favourite. Users can see when a taxi arrives and payment is made in the app by credit or debit card. Others will use public transport and apps such as **Google Maps** and **WAZE** to plan their route from home to the workplace. Updates to Google maps allow you to request wheelchair accessible routes. The augmented reality live view is also helpful for those with cognitive disabilities. Some find it easier to follow arrows or animated characters overlaid over the real world as viewed through the camera.

It is also useful to think about which technologies can be added to the tools used by your company to make some tasks easier:

EverNote is commercial software that allows you to store, retrieve, share, and collate notes and information from many sources. Other apps such as Google Keep and Microsoft OneNote are free products also available. Still, EverNote has more powerful features for business use.

Grammarly is a powerful spelling and grammar checker that can check anything you type for accuracy and ease of reading. It is a valuable way of working efficiently and effectively for many people who become anxious about their work or struggle to check their written text.

*

Accessibility in the workplace depends upon you having the right assistive technology that works with the applications that businesses use. We have introduced some of the technologies that you might need at work, and employers must anticipate your needs by striving to purchase applications that have been proven to be accessible as part of the regular procurement and planning processes.

In education

This section will discuss how technology can be an effective accommodation in the classroom. First, we will refer to universal design for learning (UDL) as supporting the individual use of technology. Next, we will focus on assistive technologies that have been proven to work well, including the accessibility features of some of the tools used in teaching and learning. Finally, we will touch upon how different technologies may be needed over time and how we consider the needs of others in the same classroom in making a decision.

We will consider both online and physical learning and refer to the different needs of different ages and different phases of education. We will outline the accessibility features of technologies that are available for people with disabilities in education, such as computers, phones, and tablets, and some of the other educational technologies that are helpful alongside commercial products that might be worth considering.

Assistive and accessible technology are increasingly essential to deliver inclusive education to meet the needs of all learners. For example, learners with limited sight can access documents if they can increase the font size, change those fonts, and adjust colour contrast settings between text and background. For those with greater challenges in engaging with text, having that text read aloud on a computer or other device offers greater ease of access.

Assistive and accessible technologies offer an effective means by which the needs of individual learners are accommodated in the classroom. Technology can support learning through accommodations to address needs by:

- facilitating access to content
- production of high quality written materials
- checking and testing of spelling and grammar
- producing evidence of learning in a variety of media
- encouraging a greater focus on learning
- increasing motivation and attention.

Accessible and assistive technologies can contribute significantly to the delivery of universal design for learning (UDL) in the inclusive classroom. The main principles of UDL are to ensure:

- multiple means of Representation – how information and resources are presented to learners
- multiple means of Action and Expression – how learners can demonstrate what they know and understand
- multiple means of Engagement – seeks to address the many ways learners are engaged and motivated within the teaching and learning process.

Accessible technologies offer a route to deliver each of these principles. For example, creating learning materials based on UDL principles is one side of a coin. The coin's reverse is assistive technologies that support or 'scaffold' the individual learner's experience.

The importance of assistive and accessible technology

Microsoft Office

Microsoft Office 365 consists of applications for email, word processing, spreadsheets, and presentations. Each has many features to make them more accessible and usable by learners with a disability. The apps integrate with assistive technologies and accessibility settings on most devices. In addition, some offer built-in Ease of Access and Learning Tools to enhance reading and writing experiences for people of all abilities. Accessibility Checkers, Accessible Templates, Autogenerated Alt-Text for images, and Captions for audio are available in the Microsoft 365 apps to make learning easier for everyone.

Some of the more useful features include:

Read Aloud reads all the text in a document using built-in voices. Many learners find this useful when reading is difficult, including temporary needs such as eye strain.

Editor offers access to Microsoft's spelling and grammar tools as well as helping to check for clarity. As with any automated checker, learners should review each suggestion as the advice is not always accurate

Immersive Reader offers a range of features to make reading easier. Changes to layout, speech output and a reading ruler can all be helpful for learners who are struggling with large documents

Focus is a useful tool for anyone who finds it difficult to concentrate on work for longer periods. Focus removes all the distractions on the computer and allows learners to concentrate on their documents.

Speech recognition is built into Office 365. It allows learners to dictate everything they want to write. This can be an alternative to writing using a keyboard all the time. Still, others find it useful when their hands become uncomfortable. It can also be a useful way of reducing the risk of RSI and strain in the hands and arms.

Accessibility Checker allows learners to check that documents and presentations will work effectively with assistive technologies. This is good for them and helps produce accessible learning resources for all.

G Suite (Google)

G Suite and Chromebooks are increasingly used in education as a low-cost means of providing access to technology for learners. Some of the key features that are included are:

- Live Captions
- Voice Typing
- Keyboard Shortcuts
- Screen Reader support
- Braille Display Support
- G Suite is compatible with several screen readers, including ChromeVox, Google's screen reader.

education

Classroom technologies that promote access

Interactive whiteboards (IWB) and voting systems – an IWB is a presentation device that links to a computer. It displays images on a touch-sensitive screen similar in size to an old-fashioned blackboard. Everything on the screen is projected to be large to easily see and manipulate. Learners can have their own tablet or similar 'pad' that they can use to contribute to what is on the screen in a way that is best suited to their needs.

Reader or Scanning pens are keys devices that a learner uses by running the 'nib' across a line of text. The text is read back aloud and displayed on the device, where each word is highlighted as it is read back. The device is small and easy to use.

Clicker 8 is a very useful means of developing literacy skills. It is based upon a friendly word processor. Learners can tackle writing tasks using speech feedback, talking spell checker and word prediction. For those just starting with reading, Clicker Grids enable them to write with whole words and phrases. Emerging writers build sentences word-by-word, while Word Banks provide scaffolding to support developing writers.

Developing literacy

Clicker Docs is a word processor that reads text aloud at each punctuation mark and offers word choices for misspelt words or from word banks you can load into the app. This app is particularly beneficial for those who struggle with written expression. It has word prediction and can help expand vocabulary choices. In addition, those who have difficulty editing their work can hear their document after inputting punctuation. This feature can help you learn to edit and revise your work independently.

Learning Upgrade is a digital solution to help adult learners accelerate growth in literacy and maths skills to succeed in classes, earn a qualification, get a better job or enter college. Learners can access over 900 lessons on smartphones or tablets anytime and anywhere. In addition, the interactive lessons feature songs, videos, games, and reward certificates.

Amrita Learning is a personalised learning app with engaging, culturally appropriate content related to life skills. The content is structured into two forms, ESL (English as a Second Language) and NS (Native Speaker). Small learning units are sequenced to build on previously learned skills within each. Specific skills used for decoding words, vocabulary, fluency, and comprehension are emphasised. The materials are designed differently but stress the use of engaging material to reduce high dropout rates.

Codex: Lost Words of Atlantis provides a different approach to learning. It combines proven teaching methods with game development expertise. The app focuses on how adults learn to read through a game suitable for all ages.

Support for maths and sciences

Calculators can help solve maths problems. Most people know about basic electronic calculators. However, many who struggle with maths prefer calculators with keys with large numbers and symbols. There are many kinds of calculators, from graphing calculators to apps. Some can solve equations with variables. For example, **Big Calculator** is an app for the iPad which turns the entire screen into a huge calculator. This can be especially helpful for those with dexterity issues or low vision.

Talking calculators look and function like common calculators. However, this assistive technology device has a built-in speech synthesizer so that each key pressed is spoken out loud. This can help the user verify that the numbers and operands have been entered

education

correctly. The calculator also speaks the answer to the maths problem. The **Talking A4 Size Desk Calculator** from Cobolt is a good example of a large key speaking calculator. In addition, **www.desmos.com/calculator** is a useful free online graphing calculator. Maths notation tools support you to write or type the symbols and numbers for equations. Writing out equations by hand can be challenging for people who have trouble writing. Most word processors are limited in handling maths symbols. **MathsType** is a maths editor for Microsoft Word. Equations are easily produced, with correct layout and symbols. This can be inserted into a Word document with a font style and size that you prefer.

Graph paper is based on a grid to make it easier to line up numbers and symbols in maths. This can be helpful for anyone who can become visually confused. Some learners prefer graph paper that has large squares. You can find examples of digital graph paper at **print-graph-paper.com**.

Graphing tools help draw the path created by an equation. If you are studying algebra or calculus, these tools can help solve graphing problems. **accessiblegraphs.org** is a good starting point. While it is not simple to use, it can be useful for anyone studying mathematics and can use a spreadsheet. Drawing tools can help to draw lines, shapes, angles, etc. Classroom tools such as rulers, stencils, and protractors can all be helpful. Still, there also are computer programs for drawing. Students studying geometry or advanced maths may find them helpful. One useful tool is **www.geogebra.org**.

Equation-solving tools are digital applications to help students work with equations. Equation-solving tools help learners find out how to solve a problem. **EquatIO** from TextHelp provides multiple ways to work with maths notation. For example, learners can type, handwrite or speak maths expressions into EquatIO. Manipulatives are objects that help us to solve maths problems in different ways. For instance, number lines help with addition or subtraction without needing to write any numbers or symbols. An abacus lets you calculate by moving beads. Physical objects are a very concrete way for many to learn mathematics. Still, for others, virtual manipulatives such as those at **oryxlearning.com/learn/7-best-virtual-manipulatives-for-online-math-learning** can be a valuable option.

Orbit Graphiti is an interactive tactile graphics display. It offers a new approach to non-visual access to graphical information such as charts, drawings, flowcharts, floorplans, images and photographs through moving pins. In addition, it features a touch interface to enable learners to 'draw' on the display. For example, tracing a shape with a finger raises the pins along the path traced.

Learning to code

Many learners with disabilities have successfully learned to code, which has opened new employment opportunities for them. The assistive technologies related to a disability and some maths tools are especially helpful as learners progress. A new initiative has been launched by Lego. This takes the idea of accessible manipulatives a stage further. It allows children to start coding by using special lego sets that link to a tablet for the first steps. More details can be found at **www.lego.com/en-gb/categories/coding-for-kids**, including sets for use at home.

Developing study skills

Part of learning is about developing study skills that make learning easier. Studying often requires the learner to store, organise, find and share information and connect pieces of

information together to form and reform new ideas. Technology can be a useful way of developing these skills where more traditional approaches are challenging. Increasingly all learners are finding these helpful.

Digital Voice Recorders

A voice recorder is a small device used to record lectures, meetings, and other important events. It is also useful for recording conversations. Such recorders are widely used and have a wide range of prices. Storage, external microphone, ease of connection to a PC and ease of use are all factors to consider when buying a voice recorder. Mainstream websites, including both Amazon and eBay, offer many for sale. It is worth reading the reviews of these carefully to see if they match your needs and circumstances. Companies such as Sony, Philips and Olympus are among some better-known brands. Some people also use apps on their phones to record speech. Apps such as **Voice Record Pro** for Apple devices and **Voice Recorder** for Android offer many features.

Organising notes and information

Evernote is commercial software that allows you to store, retrieve, share and collate notes and information from many sources. Other apps such as Google Keep and Microsoft OneNote are free products also available. Still, Evernote has more powerful features for older students.

Mindmapping through graphic thought organisers is a way of organising information. The organsiers include ways to record our thoughts and ideas and are sometimes referred to and include mindmapping. We put ideas in a 'bubble' onscreen using these tools rather than simply making a list. We then draw lines to connect each idea to others or add further thoughts and refinements to the original question or concept. Mindmaps can represent tasks, words, concepts or items linked to and arranged around a central subject. They can turn a list of information into a colourful, memorable, and well-organised diagram that may suit a learner. They are useful for planning, organising, note taking and studying.

Good starting points to introduce mind mapping are Xmind and Freemind for the PC or for more advanced features, Ayoa or **Inspiration**. A simple online tool is **www.bubbl.us** for both individuals or small groups. Tools such as **Claro Ideas** are especially helping allow ideas to be represented both by text and images.

Finding accessible content for study

Bookshare and Read2Go is a reading app for Bookshare, the online library for those who struggle with printed text. Bookshare membership is free for those who are print-impaired; downloading Read2Go lets you take your books with you on your mobile device. Read2Go can connect to braille displays via Bluetooth.

OER Commons is a public digital library of open educational resources, including books, articles, lesson materials, videos, and other types of content. All of the content is free to use and share. You can search for materials accessible to different groups of need in the advanced search option. OER commons can be accessed at **www.oercommons.org**.

*

Assistive technology is a vital part of accessible education. But the impact of that technology is much greater when the teacher uses accessible technology for all learners, applies universal design principles to the teaching, and creates and shares accessible materials. When all of these are carefully in place, a more inclusive education is possible.

Daily living

Tools for independent daily life range from essential daily activities such as bathing, feeding, dressing, and sleeping to more complex digital tools such as environmental control systems, smart homes, and alerts and alarms. These support people with disabilities interacting with their surroundings through remote-control devices, electronic doors, alternate controls, light switches, door answering or opening systems.

Independent living for people with a disability can mean very different things to different people. However, perhaps it can be defined as having the same choices and control in everyday lives that others take for granted. Technology has a vital role in achieving this by empowering people to make choices and reduce and mitigate risk. There are many ways in which assistive and accessible technologies can protect those who may be vulnerable. The growth of telecare approaches that seek to enable people with a disability and older people to live independently within their own communities is a good example.

New care solutions have been introduced, allowing users to maintain their privacy while supporting social and health needs delivered through technology. Telecare is made up of different features. Some are passive systems that monitor the well-being of people without their active involvement. Others are more interactive and require the introduction of assistive technologies to allow individuals to be active in their own care.

Monitoring and alerts

Mobile and portable technologies, including wearable devices, offer the opportunity to anticipate and respond quickly to potential problems. This can include health, activities and mood. Wearable technologies such as **Apple Watch** can alert a member of your family or a carer in the event of a fall. They can monitor your location and help find a person who has become confused or disorientated.

Fitness trackers help measure physical activity that might be required for well-being. Further prompts and alerts for daily activities, including appointments, medication or mealtimes, can reduce risks. The prompts and alerts can be received on any device, including consumer technologies.

Medication aids

There are many different medication aids available. Simple boxes for pills help people remember to take their medication on the right day and time and are often available from a local pharmacy. Automatic dispensers for pills taken regularly are also available. These are pre-filled and then locked. The dispenser sounds an alarm when you need to take your medication, and the right compartment opens. In some cases, the alarm may continue until the pills are removed. If they are not removed, some devices can alert a friend or relative to notify them.

Locator devices and solutions

These can help you find things you regularly lose, such as keys or a wallet. An electronic tag is attached to an item. Some systems come with a dedicated locator device, and if you mislay an item, you click a button on the device to make the tag beep. These systems can be confusing for some people but are helpful for carers or when carers can offer support to the person to use them. An alternative is to attach the tag to each item and link these to a smartphone using an app. A popular system is **Tile**. The system stores the last place your phone 'saw' the tile, which can be displayed on your phone.

Safety

It is important that people feel safe in their homes, especially if they live alone. Technology that supports someone to remain safe can help them to stay living independently. Technology designed to support a person's safety includes the following:

Automatic lights that are activated when someone is moving around. These help to prevent trips and falls.

Automated shut-off devices can interrupt gas supply if it has been left on or turn off a cooker. These may need to be installed, which may cost money.

Water isolation devices can turn off a tap if it's left running, preventing flooding.

Special plugs allow you to choose the water depth for a sink or bath. If it goes above that level, the plug opens, draining the water. They can include heat sensors that change the plug's colour when it reaches a certain temperature helping to prevent scalds.

Fall sensors can register if a person has fallen.

Telephone blockers can be used to stop nuisance calls.

Safer walking

People often like to walk. But for some people with a disability, there may be risks, such as the person getting lost or disorientated. As a result, some people might benefit from walking devices or apps. These include alarm systems that alert when someone has moved outside a set boundary and tracking devices or location monitoring services. These use location-based technology to locate and track the person. The types of devices include wearable, smartphone apps, keyrings and pendants. These are used when there is a risk of getting lost or disorientated. The location of the person carrying the device can be viewed remotely. Many devices also include a panic button for use if lost.

Many new mobile phones also have similar technology. These include **find my** and **Live 360** which can be used instead of a specialist device. When purchasing a device, it is important to consider how reliable it is, will it work indoors, and how often will it need charging? Safer walking technology has many benefits but raises ethical questions around capacity and consent.

Control and interaction

Technology can also facilitate real-time interactions between you and a service provider. This can be valuable when living in rural or isolated communities. Some services such as speech therapy or rehab and skills training can also be delivered through live interaction with a professional, with video and demonstrations of activities or structured applications on a computer or smartphone. Some apps can guide you in these without having a real person involved throughout. For instance, **Peak**, a 'brain training' app, can help you practise a range of memory and cognitive skills as you age. Systems to support control of the environment

Join over 100 local charitable organisations who feel better equipped to do their work and over 200 Skilled Volunteers making a difference everyday in their communities!

100%
OF VOLUNTEERS AND SOCIAL GOOD ORGANISATIONS WOULD RECOMMEND LINK UP LONDON'S SKILLED VOLUNTEERING SERVICE

98%
REPORT BEING BETTER EQUIPPED TO DO THEIR WORK

100%
FELT WELCOME AND NEEDED ON THEIR SKILLED VOLUNTEERING PROJECT

Link UP London supports small and medium sized London-based Charities, Social Enterprises and Community Groups (Social Good Organisations) who often lack specialist professional skills in-house to help them develop, grow and create a larger impact in their local community.

Many people are looking for a way to 'give back'. We connect professionals willing to share their skills and expertise with SGOs on short-term, structured projects with flexible time commitments.

We make Skilled Volunteering easy. Our supportive approach is designed to save time and resources.

www.linkuplondon.org

 twitter.com/linkuplondonuk   www.linkedin.com/company/linkuplondon

 www.instagram.com/linkuplondonuk

are changing rapidly with the advent of smart homes. Independent living may depend upon the individual's ability to control the building in which they live. Such systems can be passive or more interactive and controllable.

Passive systems may include solutions that control the environment through pre-defined patterns. For instance, heating or lighting can be set to become active at a specific time or following triggers such as a digital thermometer. These systems are influenced by artificial intelligence that can predict and anticipate needs according to the data gathered. For instance, systems can predict that heating or lighting needs to be switched on, based on a person's location as they return home. Other systems requiring action by a person with a disability are also widely available. These can be used to open and lock doors, control entertainment or curtains, and give control over the rooms of your home. These can be installed to ensure maximum independence is achieved without the need for additional hours of personal support from a carer. The growth of smart home solutions reduces the costs of home control considerably. For example, products such as **Amazon Echo** allow many people with a disability to control their environment solely through their voice. Such devices even work well for those using speech-generating devices with little or no clear speech.

Telecare

Telecare refers to systems or devices that remotely monitor people living in their homes. It enables access to support or services when needed. The technology is connected by telephone line or the internet. The systems often include alarms, sensors, movement detectors, and video calling. Telecare systems support independence and personal safety. They help reduce the risks associated with living alone and can be useful for people with disabilities. Telecare can help complete tasks such as a notification to take their medication. It may alert other people of dangerous situations like a fall or flood. The sensors around your home can link to a person or call centre. The system monitors your activities and triggers an alarm if a problem emerges. This can also be triggered by pressing a panic button or alarm.

Telecare encompasses many of the features discussed above and includes a range of sensors such as:

- community alarm such as a pendant worn by the person to press if they need assistance
- medication reminders
- flood sensors
- extreme temperatures sensors send a warning signal if the temperature is very low, very high or changes suddenly.
- absence from a bed or chair
- getting up in the night
- leaving home.
- devices to monitor daily activity.

As with tracking devices, using telecare systems poses ethical challenges.

Devices to support engagement, social participation and leisure

While assistive technology has been used to help people remain safe and continue with everyday activities, it is also used to support social activities and enjoyment. This can help

 daily living

maintain relationships, skills, and wellbeing. With the increasing availability of tablets, smartphones and apps, there are many new options to help people stay in touch and engage with those close to them. They also offer opportunities for activities, which is important for supporting the wellbeing of a person with dementia. These can include reminiscence, creative activities (such as music), video calling and life story work.

Other types of assistive technology that can be used for leisure include:

- digital photo frames – these can be programmed to show a slide show of photographs and may help support conversation with others
- puzzles and games
- sensory stimulation – devices that use touch, sound and light
- electronic games and apps
- mental stimulation, such as 'brain training' devices
- easy to use equipment such as music players and radios.

A tablet used to deliver these can become a topic of conversation and lead to more interactions for the person. This is particularly true for 'intergenerational' interactions, where the shared experience of the technology gives a younger person a connection that might otherwise not have existed.

Internet usage

Both telecare and environmental control will depend upon the ability to interact through the internet. But we have learned through the years of the Covid pandemic that the combination of assistive technology with internet friendly devices and applications can make a significant difference to the quality of life. Whilst we will not address the design of accessible websites, we note the range of applications that people with disabilities have suggested are beneficial including:

- e-commerce with home delivery
- instant messaging
- voice and video calls
- remote assistance
- collaboration and communication
- community activity
- online gaming

As services are increasingly delivered online from the 'cloud', opportunities will increase. However, enhancing life opportunities may be lost without accessible design and support for assistive technologies.

*

There are many aspects to living independently and achieving the best possible quality of life. In all its forms, technology offers people with disability access to the same opportunities and choices that others take for granted. So while we recognise some risks, we should also value the benefits that the technology offers.

Health

The use of technology to support health and wellbeing has grown considerably. Wearable technologies have demonstrated significant growth in recent years, with fitness trackers and smartwatches achieving considerable sales worldwide.

Health apps can typically be broken out into three categories:

General health and well-being apps such as nutrition-trackers help us count calories, sleep apps that track our sleep patterns, and stress-management apps that help us calm our minds.

Telemedicine apps that provide virtual patient care by licensed doctors.

Health management apps can assist individuals in monitoring their own health conditions. These can include heart disease, diabetes, pregnancy, mental health, etc. In addition, they may allow healthcare providers to share and report on a patient's personal health records remotely and help keep track of medications.

General health and wellbeing technology

Gaming for health

Wii Fit was a remarkable technology. The balance board offered several features that some people with a disability found useful. For instance, older people could practice balance and maintain posture on the board. The low-impact step activities were ideal for those who wanted a safe place to exercise. Although the Wii has been discontinued, it is easy to buy a used console online with the recommended balance board and Wii Fit Plus.

Switch – Ring Fit Adventure has been one of the most popular fitness games for a console since Wii Fit. It perhaps also redefines a little of what we mean to be accessible. In this case, the options made it possible for the game to be meaningful even if you had a physical disability. There is an Assist Mode where you can include or exclude different body parts in the game. These include Shoulder Assist, Back Assist, AB Assist and Knee Assist. By choosing Knee Assist, parts of the game that ask you to jog or run on the spot are automated. Your avatar will jog automatically through the game and sprint as needed. Those with limited use of their legs can still play the game using the resistance band to blast obstacles, collect power-ups, and fly or jump. All of the upper body exercises remain unchanged.

Amazon Echo has many fitness extensions that you can use to design a simple program for yourself. One of the more popular is **Fitness Thirty** which is a little like having a trainer in your Alexa speaker. You choose the type of exercise you want, and Alexa will call out a workout move for you to do while she counts down 25 seconds. After the 25 second countdown, she gives you a 5-second rest. Then, you can start all over again with a different move.

health

Fitness trackers

Apple Watch OS3 enhanced its health capacity with a wheelchair fitness tracker. No new hardware is needed, and the app with wheelchair fitness included is a free update. You can input different seat heights and wheel sizes. The app measures calories burned according to terrain, so you get different measurements if you are on grass or pavement. The gyroscope function measures uphill and downhill efforts, and the GPS capability can map your route. If you have a Series 1 watch, you will need to bring your iPhone along to receive GPS. Series 2 watches have an integrated GPS tracker.

Fitbit has been considered a helpful alternative as a wheelchair fitness tracker. It can count any movement as 'steps', including arm movements used to propel a wheelchair. It tracks distance, calories burned, active minutes, and sleep patterns. All of these can be monitored via the Fitbit app on your phone. Details of how to use the fitbit when using a wheelchair can be found by searching for "how do I use my Fitbit with a wheelchair' on the web.

Oura ring 3 is an interesting alternative that has been created by Oura. This is a little more expensive and works as a ring on your finger. The ring offers 24/7 heart rate monitoring, personalised health insights, and sleep analysis.

Apps

WheelFit is a workout app for Android designed for wheelchair users. It seeks to boost mental and physical health. The app provides workouts and training plans to help keep you fit.

Wheelchair Calorimeter App for iPhone allows you to use GPS coordinates to measure your location, determining when you are moving. The app uses this to estimate the energy consumed but can only show an estimated value based on your own body weight and the distance covered. In addition, your pulse rate is not measured and does not consider the ground you are on.

Other useful apps for wheelchair users include **Fit Wheelchair** and even **Fit Weightlift**.

My Fitness Pal connects to many other fitness trackers and apps to put all your fitness information into one place. As well as helping you monitor your daily activity and diet, it has many other features to help you stay healthy

Pacer monitors your movement ad activity every day. It can link to a fitness tracker or your phone. It is quite simple to use, and tracks walks, runs and rides. You can also get access to personalised fitness plans and guided video workouts.

Nutrition and diet

MyFitnessPal has an enormous food database, barcode scanner, recipe importer, restaurant logger, food insights, calorie counter, etc. It can be useful for monitoring your nutrition to lose weight, changing eating habits or simply where you want to be more careful with your health.

MyPlate Calorie Counter is an app from Livestrong that can track your nutrient intake and exercise and offers a range of recipes, meal plans, and workouts. In addition, a

community section provides peer support, motivation, and tips and tricks.

The **NHS Weight Loss Plan** app seeks to help you eat better, lose weight and get healthy. Whilst it includes the usual diet trackers, it offers a plan that is broken down into 12 weeks helping you to:

- set weight loss goals
- use the bmi calculator to customise your plan
- plan your meals
- make healthier food choices
- become more active and burn more calories
- record your activity and progress.

Telemedicine/health monitoring

Increasingly people are turning to technology to help them monitor their own health and alert them of any emerging problems. There are many different technologies available to help us to do this. Because of the complexity and range of products, it can be useful to look at this through the lens of one device. We will look at what is available if you have an iPhone. Always remember that other systems exist and may be cheaper and easier for you to use.

The **iPhone** has a health app that is the hub of all your health information. Using your phone, it can measure how many steps you are taking your walking steadiness and can link to other apps such as **Mysugr**, Diabetes Tracker Log or **Lifesum** for healthy eating

Apple Watch adds a wide range of extra sensors to gather your well-being data. It can monitor your heartbeat, log your period, record symptoms like cramps, and monitor your sleep patterns from your wrist. In addition, you can link your health records and choose if you want to share your data with a professional.

Upright Go 2 – Posture Training is a discreet trainer that attaches to your upper back and alerts you with a gentle vibration when you're slouching, helping you to achieve and maintain better posture.

Withings Thermo Smart Temporal Thermometer gives you an accurate temperature reading with a no-contact gesture while automatically syncing with the app on your iPhone or iPad. You can track temperature readings, get reminders, and enter related symptoms and medications for advice on your iPhone in the app.

Withings BPM Connect Wi-Fi Smart Blood Pressure Monitor allows you to measure blood pressure at home. It provides accurate blood pressure and heart rate scores with colour-coded feedback. BPM Connect synchronises via wifi and Bluetooth to the free Health Mate app on your iPhone or iPad. You can see your data history and share it with your doctor.

One Drop Chrome Blood Glucose Monitoring Kit is a device that ensures clinically proven accurate, reliable results in five seconds. The meter then wirelessly transmits that blood glucose data via Bluetooth to the One Drop Mobile app on your iOS device. Currently, this is only available in the US although it is possible to have it shipped to the UK.

Health management apps

The **NHS App** is a simple and secure way to access a range of NHS services on your smartphone or tablet. You can get your NHS Covid Pass, get advice about coronavirus, order repeat prescriptions, book appointments, see details of your upcoming and past appointments, get health advice, view your health records such as test results and details of your consultations and register your organ donation decision

AirMid UK is a health management app. AirMid allows you to coordinate your care across the NHS, from requesting medication to booking appointments and video consultations.

It includes:

- appointment management
- video consultations
- medication management
- your medical record
- messaging
- personal health record
- links to wearables and other apps
- allow family members and friends to look after your account if you need them to.
- customised care
- custom notifications
- service locator.

Many other apps for specific health needs such as diabetes, asthma, and mental health are available. These are frequently updated, but examples can easily be found by searching the relevant app store or marketplace.

Managing your own health has never been easier. The range of available products and apps will allow you to look after yourself better and help you be better informed when you see a doctor. Before and during an appointment, you can share data to speed up an important diagnosis. Remember that using your technology and visiting your doctor are both hugely beneficial.

Leisure

Leisure is important to our mental well-being, and shared activities offer many opportunities for social inclusion. Many tools and devices are designed to enable persons with disabilities to participate in sport and leisure pursuits – these range from adapted wheelchairs and other tools that allow participation in most popular sports. However, the growth of virtual sports, initially exemplified by the Nintendo Wii console, has encouraged increased participation in sports and leisure, both for entertainment and a healthy lifestyle.

Assistive technologies offer new opportunities for leisure and entertainment for people with a disability and contribute significantly to the quality of life. Specific areas that are worthy of further consideration include:

- access to TV and cinema
- access to books and reading
- access to gaming
- access to arts and culture
- access to music and podcasts.

TV and music

Sky, Netflix, Amazon and BBC iPlayer

Sky, Netflix, Amazon Prime and the BBC iPlayer are services that provide access to a full range of content through an Internet connection. The accessible player allows users who use assistive technologies to control their viewing through alternative controls and can be viewed with captions or audio descriptions.

Streaming services can be viewed on various devices ranging from smart TVs to computers, tablets or phones, allowing the person with a disability to have a wide range of routes to content. In addition, most streaming services have accessibility features integrated so you can activate captions or audio descriptions for individual programs wherever they are available. In some cases, the company may offer additional access features, and it is sensible to check with each.

Amazon has introduced the Amazon Fire TV Stick with Alexa Voice Remote, including voice control. This lets you search and start programmes by giving voice commands. This is in addition to the other accessibility features found in other Fire TV Stick models: Voice View, Audio Descriptions, Screen Magnifier, Text Banner and High Contrast. This principle of multiple routes to content is also very relevant to the growth of ebook readers or radio/audio content, including podcasts.

Books and reading

Amazon Kindle – it is possible to access hundreds of thousands of accessible books via

leisure

mainstream suppliers such as Amazon for Kindle. These can be read on a computer, an app for phone or tablet or ebook readers available from Amazon. Many of the books on **Bookshare** are available in braille-ready format, which can be printed or read on a braille display, whilst others are available in Daisy format. Daisy is a talking book format similar to an ePub document that can be played back on a hardware layer such as the **Victor Reader Stream** or downloaded onto a phone and played through an app such as **Dolphin EasyReader**.

Other sources of spoken content include **Audible**, which has many audiobooks for download. These are human narrated on many occasions which some people prefer to be synthesized speech. Similarly, **podcasts** and **radio** remain popular sources of content. **Qcast** is a fully accessible podcast player. Although mainstream apps for apple and Android are often fully accessible, including radio and music apps such as Spotify and BBC Sounds.

Gaming

Gradually games consoles are becoming more accessible for people with a disability.

Nintendo Switch

When launched, the system was criticised as the options contained no accessibility features. But more recently, those with low vision have welcomed the ability to zoom in on part of a screen by activating the zoom feature, and with a double press of the home button, can zoom in on the screen. In addition, you can move around the magnifying window and zoom in and out with the X and Y buttons, a handy feature.

Hori has released an accessibility controller designed for the Nintendo Switch. The Hori Flex switch and joystick interface adds different input options. It can be used with either Switch or Windows PC. In addition, gamers with a physical disability can use various devices to control the game, including single switches and some USB joysticks. They have included a remapping application for users to program six different button profiles. There is even some support for eye-tracking. Sadly, the Flex is only available in Japan at the time of going to press with this guide, and is expensive at over USD$200 for the interface.

Ring Fit Adventure

This has been one of the most popular fitness games for a console since Wii Fit. It perhaps also redefines a little of what we mean to be accessible. In this case, the options made it possible for the game to be meaningful even if you had a physical disability. There is an Assist Mode, where you can include or exclude different body parts in the game. These include Shoulder Assist, Back Assist, AB Assist, and Knee Assist. By choosing Knee Assist, parts of the game that ask you to jog or run on the spot are automated. Your avatar will jog automatically through the game and sprint as needed. Those with limited use of their legs can still play the game using the resistance band to blast obstacles, collect power-ups, and fly or jump. All the upper body exercises remain unchanged.

Other modes could be ideal for those with different needs, including older users who want to play with children or grandchildren. Other features include a sound-only version, which is great for visual impairment and those who want to exercise whilst watching TV. In addition,

the switch can be connected to a television if you want to play on a large screen, making text and graphics more comfortable to follow.

Ring Fit is an example of accessibility being baked into a game beyond technical compliance to standards. The design of these features allows you to customise to a person, using options for a range of different needs.

Steam

Steam is one of the most popular gaming platforms globally, especially those who prefer to play games on their PC rather than purchasing a dedicated console. That's an important point as PCs are full of accessibility features and options Steam can utilise. However, much of the discussion focuses on accessibility for those who are blind. This is because most blind gamers cannot access the Steam software using a screen reader – technology that reads aloud the content of screens. Without this, they cannot reach the point of browsing, downloading and playing the games.

This does not mean that no disabled people play games on Steam – it would be misleading to say otherwise! Point and Click games with limited keyboard access are highly regarded. They can be played using a range of input devices such as a touchscreen, joystick or trackball. During Covid, collaborative social games have been popular with many with disabilities to reduce a sense of isolation.

Jackbox games

One of the most popular social games on Steam is Jackbox. These are collections of mini games, which might ask you to draw a picture, type a few words or answer questions to play a range of puzzles and party games. Because the gameplay is quite simple, the instructions narrated and the graphics large and clear, they have been played by families, including older and disabled members. Jackbox games are a useful blueprint for inclusive (if not fully accessible) games. They are also ideal for co-piloting where barriers to gameplay need assistance because they are social in nature. Sharing a computer screen is inherent in the game, not a form of 'cheat mode'.

Adapted Controllers

Xbox Adaptive Controller comes with two large button pads, a D-pad, and some system control buttons. Gamers can customise it with additional parts to suit their individual needs. For example, you could connect it to the PDP One-Handed Joystick, AbleNet Switches or other add-ons.

Logitech Adaptive Gaming Kit enhances the Xbox adaptive controller with three small buttons, three large switches, two variable triggers, four light-touch switches, hook-and-loop gaming boards, and labels so you can customise them. This is an easy and cheap way of adding options to your Xbox.

Logitech Extreme 3D Pro Joystick – this joystick was not designed for the Xbox adaptive controller but it is supported. It offers a more advanced design than the One-Handed Joystick. It is a good value option for an adaptive set-up, especially if you like flight simulators.

David Banes Access and Inclusion Services

David Banes Access and Inclusion Services were founded in 2016 to support the provision and implementation of assistive technologies in communities across the world. At the heart of our approach is a breadth and depth of experience in developing policy and services to promote the inclusive use of technologies. Our services include consultancy and training to support the development of accessible technology, products, and services that meet the needs of people with a disability. We have a special interest in low-cost and open solutions and the potential impact of emerging technologies.

We seek to offer advice and guidance leading to practical steps to further the access ecosystem within a community, country, or region. This network of interrelated actions ensures that interventions and developments have the greatest impact. We have defined those components of the ecosystem as including:

Awareness	Assessment of Needs	Training
Policy	Provision of Solutions	Support
Research	Professional Development	Accessible design

Our consultancy services are highly regarded and have been used repeatedly in many parts of the world. Notably in the UK, Europe, Middle East, and Africa.

We are bound by a code of ethics which ensures that the need of the client is central to our proposal. However, we value the long-term relationships we have established over many years with clients and colleagues. Our client list includes Governments, UN agencies, disabled persons organizations, and private companies. We mentor start-ups in the field and act as a trusted intermediary between stakeholders.

If we can help you implement assistive and accessible technologies efficiently and effectively in your community reach out to us at **david@davebanesaccess.org**

You can find more about us at:

www.davebanesaccess.org

Twitter: @davebanesaccess **LinkedIn** @davebanesaccess
Facebook: www.facebook.com/accessandinclusion
Pinterest: uk.pinterest.com/Davebanesaccess
YouTube channel: davebanesaccess
Medium: davebanesaccess.medium.com
Industry News: www.accessandinclusion.news

leisure

3DRudder Foot Motion Controller allows gamers to move around with their feet, keeping their hands free. In addition, it can be used as a partial keyboard in PC gaming, taking on the role of the keyboard control keys like a joystick. This makes it a useful solution for gamers with limited use of their hands and arms.

QuadStick FPS Game Controller offers quadriplegic gamers a hands-free way to play games. You can control a game using your mouth. There are three versions of the QuadStick. Each offers a varying degree of customisation and is suited to different needs. In addition, they sell mounting kits and have Xbox Adaptive Controller support.

Optima Joystick is designed to only require limited fine motor skills. They also offer a trackball for cursor control and giant colour-coded switches. **www.oneswitch.org.uk** provides a wide range of information on accessible controllers for computers and game consoles. Other useful sites include Ablegamers and Special Effect.

Specialist games

Sometimes the complexity of modern games makes them impossible to play for some people with a disability. However, a range of games designed for screenreader or single switch users have been developed to help. **Oneswitch.org.uk** has many adapted versions of classic arcade games such as Star Wars or Pang. These are played by pressing the spacebar on a PC or a switch set up to emulate the spacebar.

There are also some great games for people who are blind. **A Blind Legend** is an audio game that can be played by blind and sighted people alike. **Blindfold Games** have developed Blindfold Racer, where players drive with their ears instead of their eyes. They have over 35 games, including cards, sports, casino games and puzzles.

Audio Games on Amazon Echo – there are so many games that can be played on Echo devices it is difficult to know where to start. Most are designed for an Echo or Echo Dot, so work by asking questions and listening to your answers. We found a few by asking 'what games can I play' of our Echo.

- Official Harry Potter Quiz
- Akinator
- Pointless
- The Magic Door
- Skyrim
- Beat the Intro.

Some games can be played by one person, but many are designed to be played with friends.

Arts and culture

Online museums and galleries

Many online art galleries and museums can be browsed online. These include the Tate Collection in London, Louvre Online from Paris, and the New York Metropolitan Museum of Art. Many galleries also offer digital tours of galleries that can be downloaded onto your smartphone to provide a mixture of written and audio information as you view exhibits. This form of inclusive gallery is increasing in many countries.

leisure

Creating art with technology often depends on having access to the same creative tools as anyone else. Programs such as Photoshop, Lightroom, Painter Indesign, After Effects, GIMP and Procreate are all widely used. Some disabled artists find that an Apple iPad Pro is a very useful starting point, whilst others use a graphics tablet with a laptop. Some artists with limited fine control, even with assistive technology, face extra challenges. For instance, some using eye-tracking find that software originally marketed for schools, such as Natural Revelation Art, is a better platform because of the larger icons and brushes than other creative packages. Many such packages include trial and demo versions, and it is often helpful to start with these or free software such as GIMP.

Music

Appreciation and Listening

Many examples of music collections can be listened to online and include huge collections through tools such as Spotify, Amazon Music, iTunes or Google Play. Increasingly such content is available via phones and tablets integrating with the onboard accessibility options for a variety of needs.

Spotify is a widely used streaming music service. It has access to many genres of music, including the latest releases and classic music from across the decades and the globe. One nice feature is the ability to create playlists of songs to reflect a theme or mood and share them with friends. Shared playlists can be discussed and edited together to create something quite unique

Composition and Performance

There has also been a growth of accessible software to create compositions in various styles. Equally, tablets and phones can be used as simple mixers of samples and tracks for performance. New interfaces such as touch, gesture and movement also open opportunities for new forms of musical instrument that are more inclusive. Three examples include:

EyeHarp is a musical instrument that only requires movement of the eyes. EyeHarp users can play in groups and orchestras together with other musicians. Playing any melody is possible, from the easiest ones, with only a few notes, to the most complex.

Arcana is a physical instrument designed on the premise that no physical disability should be a reason to miss an opportunity to play music. It is developed to meet a range of physical conditions. It can be played using different body parts such as arms, head, mouth or foot. In addition, the instrument has a range of supporting accessories, allowing every person to find their preferred way of playing.

Skoogmusic makes easy-to-play musical instruments for iPad, iPhone, and iOS devices that anyone can play. The device links to an app on a phone or tablet with Apple Music, Spotify or other music streaming service. Then, it automatically tunes your Skoog so you can play along with the songs. It may sound simple, but many find it creative and fun.

Other interesting instruments and music creation devices include **Beamz** and **Dancing Dots**. The latter is a useful tool to allow blind musicians to engage with musical notation.

*

Many leisure pursuits can be pursued using assistive technologies, and digital technologies offer new ways to participate. In this chapter, there are many that we have not mentioned, notably sports, photography, and cooking. Some additional sources of information on these are included elsewhere in the toolkit.

Mobility

Mobility is dependent not just upon wheelchairs or other aids. There is also a need for tools and devices to address mobility challenges in various ways. They include technologies that support wayfinding and orientation and provide information about the accessibility of the built environment to guide decision-making. These are increasingly built upon smartphones and tablets as consumer devices. The opportunity to travel independently and safely is much sought after by people with a disability. The classic example of this is demonstrated by those people who are blind travelling with a guide dog. These helps navigate routes and identify potential dangers. However, technology has enabled many more people to travel independently without caring for a dog in recent years.

New technologies such as SmartCanes have added audible feedback to traditional mobility canes, giving additional useful feedback to people who are blind when walking. **Microsoft Soundscape** uses 3D audio cues to enrich awareness of the word around a blind person. It helps build a mental map and make personal route choices while being more comfortable within unfamiliar spaces. Many people find navigation and finding our way around new environments challenging. This can create high levels of stress and anxiety for some people, making it difficult to travel from location to location. These technologies provide a scaffold of support to reduce stress levels and help plan and complete such journeys.

Finding people

Many jobs and activities mean being out 'in the field'. For example, if someone is working as a painter or decorator, their place of work might change every day and they may be expected to find their own way to meet with the team. In these cases, being directed to another person can be helpful. By securely sharing their location, someone can be directed to find (or be found) by following directions.

A practical application for this on Apple devices is **Find My**, where location and directions can be easily shared. In addition, if the team has a mixture of phones, including apple and Android handsets, then software such as **Life 360** can be set to share locations with specific 'circles'. This can also be helpful where a person is likely to become confused or disorientated, and the app can notify others of their location and their confusion.

Using transport

For many people who need to travel, the advent of the interface pioneered by Uber to request a taxi was very helpful. The highly graphic interface and use of GPS allow you to state where you want to travel, share your current location, and monitor the vehicle's arrival. This makes it easier to be ready on time and avoid waiting. The interface has been adopted by other companies. It is helpful when a person becomes anxious trying to describe their current or desired location to a dispatcher. Many public transport apps are available, and it is important to monitor what information is available locally. Many will include details of tickets and transport times, will alert you that a train or bus is approaching, and some may even remind you that your destination is approaching, and it is time to disembark. Those with memory and anxiety challenges can reduce travel stress using public transport.

In the UK, an app for passenger assistance for trains. **Passenger Assist** lets you request assistance from rail staff for your journey to ensure you can safely travel by train and in

comfort. This includes

- helping to navigate the station
- support when boarding the train
- meeting you from your train and taking you to your next train or the exit
- arranging a ramp on or off your train
- assistance relating to an invisible need such as autism.
- carrying your bags.

The train company you're travelling with will organise assistance for your entire journey, even if you travel by another train company as part of your journey.

Planning a route

Easily planning a route can make a great difference to the ability of a person to travel with stress and anxiety. There are now many route-finding apps that can help to do this. For example, both **Google Maps** and **Waze** offer directions for driving and cycling, walking and public transport routes. However, they are not always the easiest to use. If you plan to rely on these, some practise with a friend or family member is recommended before using them alone.

Most apps offer turn by turn navigation to guide the person each step of the way as they travel. In this way, the user only must worry about the next stage of their journey, then reorientate themselves before moving to the next. Some apps will also allow messages to be displayed on a smartwatch, suggesting how far it is to the next turning etc. These can be helpful when walking where users are concerned that constantly looking at a phone marks them out as lost and may make them vulnerable to street crime.

Following a route

Following a standard map on a device can be challenging for many people. It requires the user to orientate themselves on the map, identify and match the map to the immediate environment and then follow a route provided in text or as a line on the screen. These require a good level of information processing skills, and many individuals with intellectual challenges find this challenging or impossible.

The advent of augmented reality (AR) through a mainstream phone helps to address this. AR works by displaying digital information as a layer over what can be viewed through the phone's lens. The best-known example was Pokémon Go, which encouraged people to search out their favourite Pokémon. However, these were only visible on the phone, as a digital animation sitting in the physical world.

Google Maps AR took this concept and created something with much greater utility and value for many. In the standard Maps app, AR is activated. When the phone is held up, it recognises the immediate environment. It then draws arrows on the screen that you can follow to your destination. The AR function can also display information about the local environment, such as nearby toilets or cafes if the user selects. A similar app for Apple and Android is **Hotstepper**. Like Google Maps AR, it recognises your location and plans a route for you. However, rather than draw a line to follow, a comic character appears for you to follow along the route. For some people with cognitive disabilities, this approach is especially effective.

Finding local information

One final aspect of wayfinding is when you need to find a location for the service it provides and simply want the neatest. To some extent, both Google Maps and Waze offer some of this functionality but a third-party app such as **Around Me** may carry a lot more information and will integrate with the other apps to show routes. Around Me can be helpful when wanting to find a restaurant that caters for specific needs, finding a suitable location to meet with others or finding an ATM, pharmacy or toilet when needed. In addition, it can be a reassurance and scaffold to coping for those anxious about unfamiliar places. As with other apps, it also integrates with a smartwatch to provide information on the user's wrist rather than the phone screen for added security.

Accessibility information services

Access to information on the accessibility of locations and buildings is essential to assist people with a disability plan and engage in personal travel. Traditional publications were useful but quickly became outdated as even minor works can impact access to a building.

The advent of smartphones and apps has allowed such information to be available on demand. For example, **Wheelmap** offers both a website and mobile app that has up to date information on accessibility for wheelchair users for many cities around the world. Such applications can add user-generated data where something has changed to improve access or create unanticipated barriers

AccessAble is a website and app that tries to reduce the challenges people with a disability face. It aims to give you the accessibility information you need to determine if a place will be accessible for you. They have surveyed venues across the UK and Ireland, including shops, pubs, restaurants, cinemas, theatres, railway stations, hotels, colleges, universities, hospitals, etc. In addition, you can use AccessAble to find wheelchair-friendly venues or check out disabled access and facilities.

Finding accessible accommodation

Most major travel websites and apps have filters to search for accessible rooms. For example, Expedia, Booking.com and AirBnB all have filters that allow you only to view rooms that meet accessibility standards. In addition, some websites draw information from multiple sites. AccessibleGo.com and Bookingbility.com are two good examples.

Autonomous vehicles

In any discussion of mobility, it is nice to think ahead a little. Whilst autonomous vehicles are not yet available to drive, the first autonomous delivery vehicles are already available. This growth of autonomous vehicles will influence independent mobility for people with a disability. Early technology tests demonstrated use by people with vision loss and complex physical needs. Whilst not yet available on the market, such vehicles will reduce the need for people with a disability to use public transport or depend on others to drive. The sensors and interpretation of data that enable autonomous vehicles are also being applied to wheelchair design. This may allow users to identify a location and travel without having full control over the chair.

*

Digital tools are helping to build confidence and reduce anxiety when travelling for many people with a disability. In addition, some practice and experience can help reduce a significant barrier to access to a range of roles and jobs and reduce the cost of travel by making public transport more accessible and easier to use.

summary

Assistive technology – a summary

This toolkit offers examples and further information on almost 300 assistive and accessible technologies that can benefit people with disabilities. Regardless of any impairment or disabling condition, this toolkit seeks to offer you the first steps to achieve your aspirations. It provides initial information and links to further sources that can help you find and try technologies you may not have been aware of. The diversity of choice is breathtaking.

We have tried to suggest assistive technologies designed to address specific needs and access technologies at the heart of access to different settings. All of these tools are constantly changing and being updated. We can expect emerging technologies such as artificial intelligence, robotics, drones and 3D printing to offer even more options. Today, we are already seeing the impact of other technologies, such as the Internet of Things, Wearables, and augmented and virtual reality.

Despite this change, some important principles and approaches have remained consistent. Both Accessible and Assistive Technologies are important. Universal design and specialist design are both needed to address our needs.

The **SETT** framework, thinking about our needs, setting, tasks and technologies, provides a simple but valuable way of helping us make choices. In many cases, we can make choices ourselves about the technology that will help us. But we need a support system available when additional expertise and experience is required. We shouldn't expect to get it right the first time!

Trying different technologies is the first step to finding the right one. There are many sources of information. This toolkit can help us form the questions we need to ask to find the extra information we need.

Ultimately, the examples of technologies here in this toolkit simply provide us with examples of the types of technology currently available —and they could be updated on a daily basis. However, delving deeper can be rewarding. The examples we have chosen draw on the experience of many people with disabilities who have shared their feedback in person, at events and online. Each solution in this guide has benefitted real people, but many equally valuable technologies are not yet included. To help with this, we have included links to regular newsletters in our Further Sources of Information section on page 234 to help you keep track of the constant changes.

Assistive and accessible technologies

VISION

A Blind Legend

- App (mobile)
- FREE
- Leisure

A Blind Legend is a mobile action-adventure game with a difference.

The protagonist is Edward Blake, the famous blind knight guided by his daughter he must avoid traps and enemies on his journey through the High Castle Kingdom.

Because the knight is blind, players don't have access to video, the game is totally audio-based, which makes this an ideal game for players with a visual impairment.

Players need headphones to fully appreciate the 3D soundscape. They control the game by using multi-tactile gestures, using their smartphone's touchscreen like a joystick.

The game is co-produced by creator DOWiNO and radio station France Culture, and was partly crowdfunded.

Available at: ablindlegend.com

VISION HEARING COMMUNICATION PHYSICAL **COGNITIVE** AUTISM DYSLEXIA

AccessAble

 WEBSITE

- APP (mobile)/Content
- FREE
- Daily Living – Social – Leisure

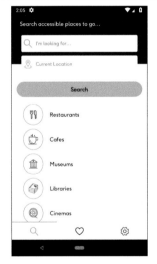
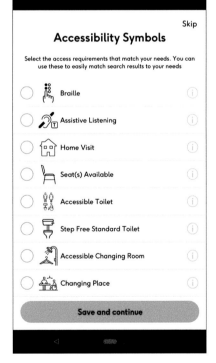

AccessAble seeks to reduce the stress and strain of planning to travel.

It provides accessibility information on venues across the UK and Ireland, including shops, pubs, restaurants, cinemas, theatres, railway stations, hotels, colleges, universities, hospitals and others.

Facts, figures and photographs help you use AccessAble to find wheelchair-friendly venues or check out disabled access and facilities.

Tens of thousands of venues across the UK and Ireland have been surveyed, and there is also a list of top tips to help you find accessible places to visit and to filter the results. Use the tag #KnowMoreGoMore.

AccessAble was formerly known as DisabledGo.

Available at: www.accessable.co.uk

VISION HEARING COMMUNICATION PHYSICAL COGNITIVE AUTISM DYSLEXIA

AccessNow

   WEBSITE

- APP (mobile)/Content
- FREE
- Daily living – Social – Leisure

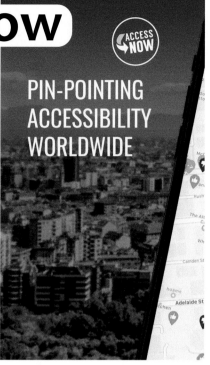

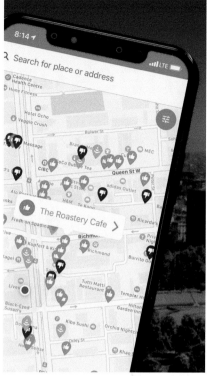

AccessNow is an app and website that lets you find, review and share accessible locations worldwide.

Every review submitted helps to build an extensive database of accessible places.

As well as contributing reviews, you can search to find sites with the accessibility features you want, rate places you have visited to keep the reviews updated, and discover accessible places nearby and globally.

The concept of a connected world community that helps to remove barriers is widened by the fact that businesses can get verified on the site to claim their accessibility listing on the app. They can manage their pages for free or pay to upgrade features.

Information on how to contribute can be found at: accessnow.com/getinvolved.

Available at: accessnow.com

VISION HEARING COMMUNICATION PHYSICAL COGNITIVE AUTISM DYSLEXIA

Access Rating

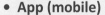

- App (mobile)
- FREE
- Daily living – Leisure

The Access Rating app is a user-led collection of reviews intended to rate venues on their disabled access.

It covers over 100,000 restaurants, bars, and hotels across the UK and users can easily search for a specific venue to read or submit a review.

The app also features an interactive Google Maps to help users locate venues.

Existing reviews can be up or downvoted and there are options to suggest a new venue or to report listings as incorrect.

Info at: www.accessrating.com

75

VISION HEARING COMMUNICATION PHYSICAL **COGNITIVE AUTISM DYSLEXIA**

Accessibility Checker

 WEBSITE

- Software
- FREE
- Daily Living – Employment – Education

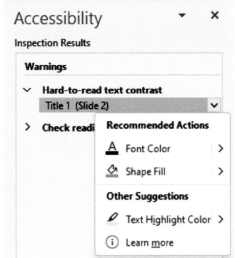

The Accessibility Checker is a tool built into Microsoft Office apps that helps to ensure that your documents are accessible.

You can use it in Word, Excel, Outlook, OneNote and PowerPoint on Windows, Office for the web, Visio and Mac systems.

The checker finds most accessibility issues and explains why they might be a problem for someone with a disability. It also offers suggestions on how to resolve each issue.

Although the Accessibility Checker catches most types of accessibility issues, there are some issues it is not able to detect – so it is important to personally check documents as well.

Available at: support.microsoft.com

VISION HEARING COMMUNICATION PHYSICAL **COGNITIVE AUTISM** DYSLEXIA

Action Blocks

- APP (mobile)/Integrated or download from Google Play operating system
- FREE
- Daily Living – Employment – Education – Social – Leisure

Android Action Blocks are a useful tool for users with a cognitive disability and assists them in using an Android device to carry out routine actions.

Each Action Block represents an action such as making a phone call, or switching on the lights and when activated, it will trigger the Google Assistant to perform that action.

When setting up Action Blocks there is a list of common actions to choose from. Each Action Block created can be given either a name, an image/photo, or both. The user can also opt for the device to vibrate and/or speak action aloud when an Action Block is pressed. Once created, the Action Block can be placed on the user's home screen, and they simply need to select it to perform the action.

If required, Action Blocks can be linked to adaptive switches.

The Action Blocks app is currently available in eight languages, including English.

Available at: play.google.com

PHYSICAL

Adaptive game controllers

- Hardware/Consumer tech
- ££/£££
- Social – Leisure

Adaptive game controllers are hardware devices that replace standard controllers provided for a computer or console with a version that is specifically designed to accommodate different body functions or movements

Some adaptive game controllers are mainstream technologies such as the

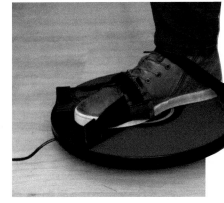

3dRudder Foot Motion Controller (pictured) which whilst designed as a gaming accessory, can also be used as your main control. Other controllers allow you to attach switches, joysticks or trackballs to control your game through a different device.

There are many useful sources of information regarding controllers given below:

www.thechildrenstrust.org.uk/brain-injury-information/latest/xboxs-adaptive-controller-a-game-changer-for-people-with-disabilities?

lifezest.co/adaptive-gaming-controllers/?utm_source=rss&utm_medium=rss&utm_campaign=adaptive-gaming-controllers

www.spinalcord.com/blog/best-adapted-video-game-controllers

www.pcmag.com/picks/gaming-for-everyone-6-accessible-gaming-devices

HEARING

Agrippa Deaf Alert Pillow Fire Alarm

- Hardware
- £££
- Daily living

The Agrippa Deaf Alert Pillow Fire Alarm consists of a battery-operated unit that wirelessly connects to a pad placed under the user's pillow.

It is designed as an alert system for users who have a hearing impairment that would prevent them from

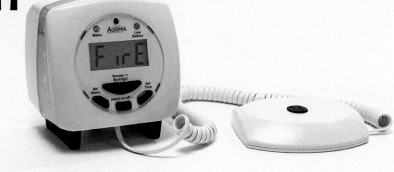

hearing an audible fire alarm. If the pillow alarm is triggered unit will display bright red, LED lights and the word 'Fire'. Meanwhile, the pad will vibrate and wake the user.

The Agrippa Pillow Fire Alarm uses 'listen-and-learn' technology that learns to identify the sound of a specific fire alarm. It can also function as an alarm clock.

Info at: www.geofire.co.uk

VISION HEARING COMMUNICATION PHYSICAL **COGNITIVE AUTISM DYSLEXIA**

Airmid UK

- App (mobile)
- FREE
- Daily living

Airmid UK is designed to help users organise their medical information within one app. Users can log into Airmid using their existing system online username and password. Alternatively, they can use their NHS login details.

Once logged in, users can access information such as their medical notes, medication and vaccination history, allergies and drug sensitivity information, and any medical letters or reports. They can also use the app to book and manage online appointments or request repeat medication.

Airmid UK provides a link between users and their GP practice or other medical organisations that care for them. It will also upload data from their own personal health technology such as Apple Health or Google Fit.

Other features include:
- managing preferred pharmacy and map details
- using Electronic Prescription Service, providing barcodes for medication collection
- direct messaging to contact a user's GP/clinician
- keeping records of communication with medical practitioners
- setting medication reminders.

Available at: airmidcares.co.uk

COMMUNICATION PHYSICAL

AMAneo Assistive Anti-Tremor Mouse

   WEBSITE

- Hardware
- £££
- Daily Living – Employment – Education – Social – Leisure

The AMAneo Assistive Anti-Tremor Mouse Adapter can be connected to any mouse.

Depending on the type of mouse being used, the adaptor electronically filters out – and compensates for – any trembling of the user's hand, head or limbs. This allows the mouse pointer to move smoothly on the screen.

The adapter works with any mouse and any operating system. No software installation is needed. The adapter is simply plugged in between the mouse and PC or another device via USB.

It is easily adapted for different users, and the intensity of the tremor filter can be adjusted according to user needs. It also has slots that allow you to connect separate switches for the left and right mouse buttons.

Available at: Adapt-IT.co.uk

VISION HEARING COMMUNICATION PHYSICAL **COGNITIVE AUTISM DYSLEXIA**

Amazon Alexa

 WEBSITE

- Hardware – App (mobile) – Consumer tech
- App: FREE
  Echo Dot: ££
  Other accessories: £/££/£££
- Daily living – Employment – Education– Social – Leisure

Amazon Alexa, (often known simply as Alexa), is a, primarily voice-controlled, virtual assistant.

It can be accessed/controlled by use of the Alexa Smart Speaker, although this has to a great extent been replaced by the smaller-sized Echo Dot speaker.

It can also be controlled by the use of the Alexa App. It uses a combination of AI, automatic voice recognition, and natural language processing to implement a user's commands.

Users can activate Alexa by use of a wake word such as 'Alexa' or 'Amazon' (there is an option in settings to change the wake word). However, when using the app, it may be necessary to click a button.

Alexa is capable of voice interaction, setting alarms, making to-do lists, managing shopping lists, streaming podcasts, playing music and radio, playing audiobooks, and providing real-time information such as news updates, and weather/traffic/sports reports. Also, users can add to Alexa's capabilities by installing a range of additional skills, such as using a different language or suggesting recipe ideas.

Alexa can also control several smart devices functioning as a home automation system. This enables users to control their environment, e.g. remotely operating lighting or heating systems. Several Echo Dots can be positioned and connected to provide whole-home coverage. Other hardware, such as cameras can also be added to the system.

This remote, hands-free method of controlling the environment around us can be especially helpful for people with physical disabilities who want control over their homes, but it is equally helpful for people with vision loss or other disabilities where using speech or using icons on a phone feel more natural.

Additional skills suggestions:
www.tomsguide.com/uk/round-up/best-alexa-skills

Many instructional videos are available to demonstrate the range of skills/services, e.g. Amazon Alexa – A Complete Beginners Guide: www.youtube.com/watch?v=6mSuxMfJZMs

Available at: www.amazon.com

VISION HEARING COMMUNICATION PHYSICAL COGNITIVE AUTISM DYSLEXIA

Amazon Fire TV Stick

- Consumer tech
- ££
- Leisure

Amazon Fire TV Stick is a portable streaming device that plugs into a TV's HDMI port to allow users to access streamed TV channels such as Amazon Prime and Netflix.

The Fire TV Stick comes with a remote control that is used for operating the television, navigating onscreen menus, etc. It comes in a range of models from Fire TV Stick Lite (which streams shows in HD, but a basic remote that won't operate things like TV volume) up to Fire Stick 4K Max (which streams in 4K and supports the fast Wi-Fi 6).

The Amazon Fire TV Stick has several accessibility features and supports Closed Captions where available. Features include:

Voice View – a screen reader that leads onscreen text aloud to help the user navigate menu options.

Audio Descriptions – provide spoken narration for compatible shows. (Describes characters, scene changes, actions, onscreen text, etc.

Screen Magnifier – will zoom in/out to magnify the screen menus etc (not compatible with most video content).

Text Banner – creates a customisable onscreen display box that contains information such as the title of the show currently playing

High Contrast – changes to colour of onscreen text to black or white to provide a higher visual contrast.

See also Amazon Fire TV Stick.

Available at: www.amazon.co.uk

VISION HEARING COMMUNICATION PHYSICAL COGNITIVE AUTISM DYSLEXIA

Amazon Fire TV Stick with Alexa Voice Remote

 WEBSITE

- Consumer tech
- ££
- Leisure

The Amazon Fire TV Stick with Alexa Voice Remote does everything the other Amazon TV fire sticks do, with the added bonus of voice control.

The Alexa voice remote enables users to search and launch programmes simply by giving voice commands.

The voice control feature is accessed by pressing and holding the Voice button.

Television control features such as power, volume, and mute can be accessed by buttons on the control.

The Amazon Fire TV Stick with Alexa Voice Remote shares the same accessibility features found in all the other Fire TV Stick models: Voice View, Audio Descriptions, Screen Magnifier, Text Banner, and High Contrast.

See also Amazon Fire TV Stick.

Available at: www.amazon.co.uk

VISION HEARING COMMUNICATION PHYSICAL **COGNITIVE** AUTISM **DYSLEXIA**

Amazon Kindle

- Hardware/consumer tech
- ££/£££
- Education – Leisure – Employment – Daily life

Amazon's Kindle is a range of e-readers that enable you to browse, buy, download and read e-books, newspapers, magazines and other digital media wirelessly at the Kindle Store.

You can read your own books and documents copied from your computer or emailed to your device directly. Since it was launched, Amazon has developed the accessibility of its devices and apps for phones and tablets.

Those with low vision or other reading disabilities can make reading easier by customising their reading environment to better meet their needs. This includes helping to focus on individual lines of text using a reading ruler.

There is no screen glare, even in bright sunlight on a Kindle. Some devices have a built-in light. If you use assistive technology such as screen readers or refreshable braille displays, they can be used to read with Kindle apps on Fire Tablet, iOS, Android or Kindle for PC. Kindle devices rely on Bluetooth to connect to speakers, headphones and enable the on-device screen reader.

You can navigate your library or within a book by touch using VoiceView gestures on Kindle devices and your Fire tablet or common accessibility gestures on VoiceOver, Talkback, NVDA or JAWS.

Available at: www.amazon.co.uk

VISION HEARING

Amplicomms Extra Loud Vibrating Clock

- Hardware
- ££
- Daily living

The Amplicomms Extra Loud Vibrating Clock (TCL350) is a radio-controlled alarm clock with automatic time setting. It is a useful device for users with hearing or visual impairments.

The alarm comes with 3 settings, the loudest of which is 80dB. Users can also connect a vibrating pad (included) which can be placed under their pillow to provide a physical alert.

The blue backlit display has extra-large digits (3.6cm), and an ultra-bright visual alert when the alarm sounds. It has a radio-controlled feature that can be used to set the time automatically.

Other features include:
- Alarm buzzer with 3 different frequencies.
- Switchable time display: 24 hours or 12 hours.
- Automatically changes for British Summertime, etc.

Available at: www.amplicommsonline.co.uk

HEARING

Amplified TV headset

- Consumer/Hardware
- ££/£££
- Daily Living – Leisure

An amplified TV headset is intended for users who don't need hearing aids all the time but perhaps have a mild hearing impairment that makes it difficult for them to hear a television.

They consist of a wireless headset that connects to the television via a transmitting base. The transmitter itself usually plugs into the television's headphone jack.

The user can hear the television through the headset and adjust the volume and tone of the sounds they hear.

An amplified TV headset is very useful when more than one person is watching television. The user can hear the sound without the volume of the television having to be increased for the other watchers in the room.

Also, several headsets can be linked to one transmitter so multiple users can watch and hear the same television. Because each headset has its own controls, every user can adjust the sound to suit their individual needs and preferences.

Many amplified headset transmitters use an infrared signal that requires a clear line of sight between the headset and the transmitter base. Some transmitters, however, give a strong signal that can still be picked up even if the headset user moves to a different room.

Amplified TV Headsets are available across a range of brands and prices.

VISION HEARING COMMUNICATION PHYSICAL COGNITIVE AUTISM

Android Accessibility Suite

- Integrated
- FREE
- Daily Living – Employment – Education – Social – Leisure

The Accessibility Suite includes a collection of accessibility apps that enable users to control their Android device either with a switch or eyes-free.

Accessibility Suite Apps include:

● Accessibility Menu: a large onscreen menu offering users the usual options to take screenshots, lock the phone, control volume/brightness, etc.

● Switch Access: instead of using the touchscreen, users can interact with their Android device using a keyboard or a range of switches (switches include those that respond to physical pressure and Camera Switches controlled by facial gestures).

● TalkBack screen reader: gives users spoken feedback and allows them to control their device with gestures or typing with the onscreen braille keyboard.

● Select to Speak: users can select items on their screen and hear them read aloud.

Info at: support.google.com

COMMUNICATION PHYSICAL

Android Camera Switch

- Integrated
- FREE (integrated into Android devices)
- Daily Living – Employment – Education – Social – Leisure

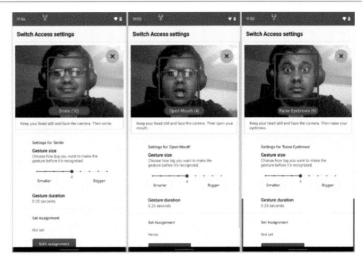

Camera Switch is designed to make Android phones more accessible to users with motor impairments and/or barriers to speech.

It enables users to turn their phone's front-facing camera into a switch operated by facial gestures.

Thus, they can navigate their phone and give instructions such as next, scan, pause, select, by eye movements and expressions alone.

The app is fully customisable to a user's needs, abilities and preferences, using gestures such as looking right/left, big/small smile, raising an eyebrow.

Users record the gestures on their phones and choose which actions to ascribe to gestures or combinations of gestures. Duration of a gesture can be set before it triggers an action.

Info at: support.google.com

VISION

Android Magnification

- Integrated
- FREE
- Daily Living – Employment – Education – Social – Leisure

The Magnification feature on Android devices is designed to allow users to zoom in or magnify the screen up to 8x to see/read it more easily. This is a useful tool for users with a visual impairment.

The Magnification feature is accessed via the Accessibility button.

Users can tap the screen and then use a two-finger pinch gesture to adjust the zoom. The screen will stay zoomed until the user switches off magnification or automatically zoom out when the user closes or opens an app.

Alternatively, the user can zoom in temporarily by touching and holding the screen rather than tapping.

Info at: support.google.com

VISION PHYSICAL

Android Switch Access

- Integrated
- FREE (integrated into Android devices)
- Daily Living – Employment – Education – Social – Leisure

Android Switch Access enables users to interact with their Android device by using adaptive switches or an external keyboard instead of the touchscreen.

Switch Access highlights each item on the screen until a switch is activated to select whatever is currently highlighted. It is a useful feature for users with limited mobility and/or face barriers to using fine motor skills to control the touchscreen.

On first using Switch Access, the user can ascribe actions to specific switches or keyboard keys according to their preference. These can be changed at any time. They also choose from several types of screen scanning (auto-scan, point scanning, linear scanning, row-column scanning and group selection. If using Auto-scan, there are options to adjust features such as the scanning speed and how often it will repeat scan a screen.

If the user has a visual impairment, there are speech sound and vibration settings that allow them to receive verbal or physical feedback about the scanning actions on the screen.

Info at: support.google.com

VISION

Android TalkBack

- Integrated
- FREE
- Daily Living – Employment – Education – Social – Leisure

TalkBack is a screen reader built into the Android operating system on phones and tablets. It is used to speak aloud text and image descriptions on the screen. It

enables users with a visual impairment to navigate their device and access printed materials such as emails and websites.

TalkBack can be initiated in several ways, either through your device settings, with a volume key shortcut, or by voice (through Google Assistant). Once open, it can be operated through touchscreen swipes and finger gestures.

A TalkBack Braille Keyboard setting is also available. (Note that the braille keyboard is not compatible with Google Docs.)

On first use of TalkBack, a TalkBack Tutorial opens automatically to guide the user through how to access all the features. Users can customise settings to suit their preferences, e.g. command gestures, menus, voice, narration speed, language (it currently supports over 60 languages.)

Info at: support.google.com

VISION HEARING COMMUNICATION PHYSICAL **COGNITIVE AUTISM**

Apollo Ensemble

- Software
- £££
- Social – Education – Leisure

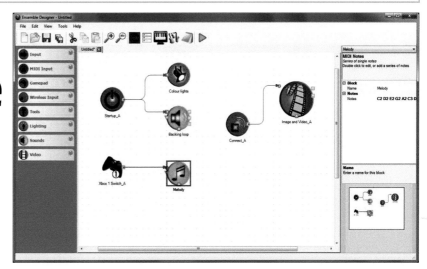

Apollo Ensemble is Windows software used to link together inputs from movement sensors and switches to various outputs.

It can create sensory environments, interactive spaces and accessible musical instruments for users with a wide range of abilities and disabilities.

Easy to use with simple drag and drop, Apollo Ensemble works with a wide range of adaptive controllers. It uses inputs such as trackballs, smart devices and games controllers.

Outputs can include videos, lighting technology and audio effects.

Available at: www.apolloensemble.co.uk

PHYSICAL

Apple Assistive Touch Draft

- Integrated
- FREE (= iPhone/iPad feature)
- Daily living – Employment – Education – Social – Leisure

Apple's Assistive Touch feature is designed to aid users with physical needs by enabling them to adapt their device touchscreen to reflect their preferences. If a user finds it difficult to perform certain gestures, such as taps or pinches, they can substitute them with a preferred gesture.

The Assistive Touch feature is fully customisable to individual needs and preferences. The buttons in the menu can be re-ordered and be assigned different content. The user can create new gestures and assign custom actions to manoeuvres, such as taps and presses.

It is also possible to connect an external device such as a trackpad or Bluetooth mouse to control the onscreen pointer. Apple Watch users can use the Assistive Touch feature to control their watch by arm movements and fist clenches.

Available on: iPhone, iPad, Apple Watch, iPod Touch.

Info at: www.apple.com

British Assistive Technology Association

Our aims

○ to campaign for the rights and interests of those needing AT

○ to provide expert and impartial support and advice to government departments and agencies

○ to educate and inform widely on the benefits of AT

BATA members are suppliers, Assistive Technology (AT) professionals and organisations that provide support to individuals with disabilities who need AT solutions. We provide an information and signposting service to the public by directing enquiries to relevant members (and other relevant groups/organisations).

On behalf of our members, we represent AT interest to the government as a grassroots organisation. BATA is a not-for-profit, the membership fees pay for our executive director and our council members are all volunteers and serve a term of office.

You can play a part too! If you are a supplier or provider of AT services then please consider joining us. We represent our members interests to the government and seek to improve the understanding of AT in the UK.

Join us in making a difference to many lives through the use of Assistive Technology.

bataonline.org

VISION PHYSICAL COGNITIVE AUTISM DYSLEXIA

Apple Find My

 WEBSITE

- Integrated
- FREE
- Daily Living – Employment – Education – Social – Leisure

The Apple Find My feature can be used to locate lost devices and share a user's location. Apple devices can be linked electronically, whilst other items can have an Apple Air Tag attached.

When searching for a linked device, users can prompt it to play a sound. If the sound cannot be heard (or if searching for an item with an Air Tag), the user can view the device's location pinpointed on a map. (If the device is not currently connected to the Internet, it will show the last known location).

If Find Me cannot locate a device, the user can mark it as 'Lost'. Doing this will remotely lock the missing device with a passcode. Users can also choose to display a custom message with their phone number on the lock screen so a finder can contact them.

Apple users can use Find Me to locate any other users who have shared their location. Similarly, if a user chooses to Share My Location with someone else, that person can use Find Me to locate them. Available on: iPhone, iPad, Mac, iPod Touch.

Info at: www.apple.com

Apple Guided Access

- Inegrated
- FREE
- Daily living – Employment – Education – Leisure

Apple's Guided Access feature can temporarily restrict device functions or access to apps. This assists users who find it difficult to stay focused on one task. It is fully customisable to suit an individual user.

When it is switched on, Guided Access will disable the Home Button and limit a device to access one app at a time. It can also set limits as to which app features are available. By turning off the keyboard and areas of the screen that are not relevant to a task, a user is protected from distractions caused by a finger gesture accidentally turning something on.

Customisable aspects include setting a time limit for using an app and/or how long before auto-lock engages during a Guided Access session. Users can choose from a list of functions to disable, such as volume buttons and motion response, e.g. when turning the device, switches the screen from portrait to landscape.

Available on: iPhone, iPad, iPod Touch.

Info at: www.apple.com

87

HEARING

Apple Live Listen

- Integrated
- FREE
- Daily Living – Employment – Education – Social

The Apple Live Listen feature is designed to assist users in holding a conversation in a noisy environment.
Users who have a hearing impairment may find it difficult to fully participate in conversations in busy, loud places.
If users find it difficult to hear

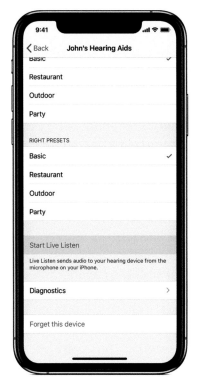

what is being said or distinguish from background noise, they can use Live Listen to hear the speaker more clearly through their wireless headphones.

Once Live Listen is switched on, users should move or angle their device towards another person. The microphone will pick up the audio and send it to their headphones. The speaker does not have to be very close by, as Live Listen will pick up speech from up to 15m (almost 50 ft) away.

So as well as being a useful feature for chatting with friends, it can also benefit a work or educational environment.

Available on: iPhone, iPad, iPod touch, AirPods Max, AirPods Pro, Powerbeats Pro, Made for iPhone hearing aids.

Info at: www.apple.com

VISION

Apple Magnifier

  | WEBSITE

- Integrated
- FREE
- Daily Living – Employment – Education – Social – Leisure

The Apple Magnifier is an in-built feature on iPhones and iPads that uses the camera as a digital magnifying glass.
When a user directs the camera at an object or a piece of text, Apple Magnifier will

work to display an enlarged image of the subject on the user's screen.

On the latest iPhones/iPads, Apple Magnifier offers a People Detection mode. It uses light reflection technology to detect when people are nearby and notifies the user via speech (Voiceover), sounds or haptic feedback.

Info at: www.apple.com

HEARING

Apple Sensory Alerts

- Integrated
- FREE
- Daily Living – Employment – Education – Social

The Sensory Alerts feature allows Apple users to choose how they wish to be notified of incoming calls, new texts, new and sent emails and other events.

There is a choice of audio alerts, vibrations or a quick LED Flash.

Apple Watch users can turn on the device's Taptic Engine to receive a gentle tap whenever they get a notification.

For iPhone users, there is the option for their screen to display a photo of the caller. Mac users will see their screen flash to notify them that an app needs their attention.

Apple's Sensory Alerts feature could be handy for a user with a hearing impairment or other reasons that they might not notice the standard message alert sounds.

Available on: iPhone, iPad, Mac, Apple Watch, iPod Touch.

Info at: www.apple.com

HEARING COMMUNICATION

Apple SignTime

 WEBSITE

- Software – Consumer tech
- FREE
- Daily Living

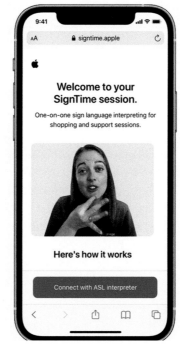

Apple SignTime is a service provided by Apple to enable its customers to communicate with AppleCare and Retail Customer Care by using British Sign Language (BSL).

Apple customers who wish to use the SignTime service should go to the SignTime webpage and click 'Connect with a BSL Interpreter'.

They will then need to give their browser permission to access their device's camera and microphone before connecting to an interpreter via video link.

SignTime users can converse with the interpreter in BSL. The interpreter will contact Apple Support on the user's behalf and translate for them. After the call, the user will receive a brief survey asking about their experience with the interpreter.

If a customer is visiting an Apple Store in person, they can also use SignTime to access a sign language interpreter to help them communicate with the in-store staff.

It is not necessary to book this service in advance.

Info at: www.signtime.apple

> HEARING

Apple Sound Recognition

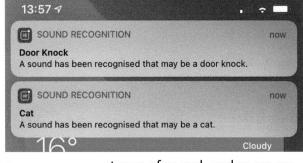

- Integrated
- FREE
- Daily living

The Sound Recognition feature on Apple devices uses on-device intelligence to listen out for specific sounds and then send an alert to notify the user.

This is a useful feature for users with hearing impairments as they can receive both visual and a physical (vibration) warning.

It is also useful if the user finds it difficult to pick out individual noises over background noise, such as when listening to music.

The Sound Recognition feature can detect a variety of types of sound, and users can choose which sounds to detect when they set up the feature. It can be used to notify a problematic situation, e.g. an alert about the sound of running water can warn the user of a tap being left on.

Available on: iPhone, iPad. Alerts will also appear on Apple Watch.

Info at: www.apple.com

> PHYSICAL

Apple Switch Control

- Integrated
- FREE
- Daily Living – Employment – Education – Social – Leisure

Apple Switch Control is an accessibility feature that allows users to control their Apple devices in various ways. It encompasses adaptive switch hardware, wireless game controllers, or even just simple sounds.

It works with various Bluetooth-enabled devices, such as joysticks, tracker balls, switches, a keyboard space bar.

A valuable feature for users with limited mobility is that it offers three ways to navigate the Dock, onscreen menus and keyboards: point scanning, item scanning, and manual scanning. Users also can create their own custom keyboards and panels, either system-wide or app by app.

If users have synced multiple devices with their iCloud account, they can use Platform Switching to use a single switch etc, to control all their devices.

With Switch Control, users can use their adaptive switch to select, tap and drag items on the screen. The user employs the switch to select an item or location on the screen, then uses it to choose an action.

Available on: iPhone, iPad, Mac, Apple TV, iPod Touch.

Info at: www.apple.com

> PHYSICAL

Apple Touch Accommodations

- Integrated
- FREE (feature of iPhone/iPad)
- Daily Living – Employment – Education – Social – Leisure

Apple Touch Accommodations feature help if you find it difficult to accurately perform actions that require quick taps onscreen or when tapping and holding. the feature allows users to adjust device settings to change how the screen responds to their touch.

The settings affect how long a user needs to touch the screen before recognising it as deliberate action. There is also the option to decide if repeated touches will be ignored.

The Touch Accommodations setting is customisable, and the touch timer can be set between 0.1–4 seconds. This feature is a particularly useful tool for users with limited fine motor control or who have conditions such as hand tremors.

Available on iPhone, iPad, Apple Watch, HomePod, iPod Touch.

Info at: www.apple.com

> VISION COGNITIVE DYSLEXIA

Apple VoiceOver

- Integrated
- FREE
- Daily Living – Employment – Education – Social – Leisure

Apple VoiceOver is an inbuilt screen reader on Apple devices and Mac computers.

It can describe everything on the user's screen: text, objects and people. For example, it can read an email or describe a photo with simple phrases such as "black dog swimming in a lake".

Spoken descriptions of actions can also help users navigate their screen using touchscreen gestures, a trackpad or a Bluetooth keyboard. Apple VoiceOver is also compatible with refreshable braille readers, so users can receive a braille description of what is on their screen.

VoiceOver does not simply read/describe the text/images on a screen. It can audibly feed information back to the user about anything on the screen. For example, it can report battery level, which orientation the screen is using, if the lock screen is on, etc.

VoiceOver on a touchscreen device is controlled by a customised series of gestures and taps. When a user drags their finger across the screen, VoiceOver announces the name of any item it is touching.

Apple VoiceOver supports a wide range of iOS apps such as email apps, Dropbox, map readers and many more

Available on: iPhone, iPad, Mac, Apple Watch, Apple TV, HomePod, iPod Touch.

Info at: www.apple.com

VISION HEARING COMMUNICATION PHYSICAL COGNITIVE AUTISM DYSLEXIA

Apple Watch

- Hardware/Consumer tech
- £££
- Daily living – Leisure

Apple Watch is a range of smartwatches from Apple. They incorporate fitness and health tracking, communication, notifications and alerts, and can control other devices.

The Apple Watch operates mostly with your iPhone, where you can configure your watch and sync information. More recent versions of the Apple Watch will allow you to text, make calls, and listen to music. Those models that connect to phone networks also allow you to stream audio or other content without having your phone to hand.

The Apple Watch has many features to help monitor health and well-being, including mindfulness and mental health apps. Other features include a heart rate monitor, sleep monitoring, fall detection, blood oxygen, and ECG.

The latest versions allow you to type messages on your phone, stream music, and use your phone as a payment system at restaurants, etc. Most recently, as a 'key' for your home, if you have compatible locks on your doors.

Available at: www.apple.com

VISION

Apple Zoom

- Integrated
- FREE
- Daily Living – Employment – Education – Social – Leisure

Apple Zoom is designed to enlarge everything on a user's screen to make it easier to see/read. Users can set the level according to their needs, up to 15x.magnification (default is set at 5x).

Users can choose full-screen magnification or a picture-in-the picture view where only a desired section of the screen is magnified. This part-screen magnification 'lens' can be dragged around the screen to enlarge areas. Mac users have the option of Zoom Display which allows them to simultaneously see content up close and at a distance, with one display zoomed in and the other at standard resolution.

Apple Zoom has customisable features. As well as adjusting both the magnification level and the size of the lens, users can

apply filters such as inverted contrast, greyscale/ inverted greyscale and low light.

Zoom can also be paired with VoiceOver as an additional aid for users with visual impairment.

Available on: iPhone, iPad, Mac, Apple Watch, iPod touch.

Info at: www.apple.com

PHYSICAL COGNITIVE

Arcana Strum

- Hardware/Consumer tech
- £££
- Social – Leisure

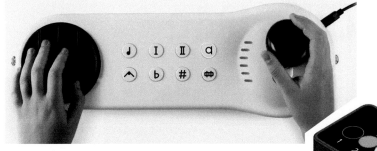

Arcana developed the Arcana Strum to make playing an instrument accessible to users whatever their age, ability or disabilities.

It is designed to be similar to a guitar. It can be played by moving the handle in strumming movements with one hand while the other hand presses on keys to emulate pressing on strings to produce chords. To create music, the instrument then needs to be connected to an external output such as an iPhone or iPad.

Arcana sells a range of accessories to complement the Arcana Strum and increase its accessibility to people with various physical and cognitive disabilities. One, the E-box, allows up to five adapted switches to be connected to control the Arcana Strum – pressure switches, motion sensors, light sensors, etc.

The instrument can also be adapted by adding one of the alternative handles. Another adaptation is to simplify the playing experience by changing the 5-key keyboards to a 3-key keyboard.

The Arcana strum can be used in conjunction with various music-making apps, such as GarageBand or Piano MIDI Legend.

Info at: arcanainstruments.com

VISION

Ariadne GPS

- APP (mobile)
- £
- Daily Living

The Ariadne GPS app offers many features found in similar GPS apps. However, one particular aspect of it makes this app a useful tool for people with a range of visual impairments.

The app uses sound, vibration and voice alerts to help users navigate their way and contributes to a level of independence and freedom in their daily lives.

By enabling the VoiceOver setting on their phone, users can access a range of information about where they are or where they are about to be.

Know your position: tapping the map will give a spoken confirmation of the user's location

Exploring the map: Users can move their fingers across the map to hear street names and numbers being spoken aloud

Alerts: Users can save their favourite points, for example, a particular bus stop, and receive a warning when they are close by. (Users have a choice of type of alert – vibration, sound or voice)

The Ariadne GPS app is currently available/localised in 14 languages.

Info at: www.ariadnegps.eu

GENERAL

AroundMe

  WEBSITE

- App (mobile)/Consumer
- FREE
- Daily Living – Social – Leisure

AroundMe is a mobile application for iOS and Android that allows you to find nearby points of interest such as restaurants, hotels, theatres, parking and hospitals.

It also provides extra information on locations such as film showings or menus.

AroundMe offers an Apple Watch app to display results on your wrist while your phone is safely in your pocket.

Once you find a place you can use the app to navigate to the venue directly within the app or by transferring the result to Apple Maps or Google Maps or another GPS system.

The app is available in English, German, French, Spanish, Italian, Japanese, and Chinese.

Available at: aroundmeapp.com

COGNITIVE   AUTISM   DYSLEXIA

ATracker

  WEBSITE

- App (mobile)
- FREE/£/££
- Daily Living – Employment – Education – Social

Pie chart, bar chart and data export

ATracker is a time tracking application that is very easy to use and requires minimal setup.

On the main screen, you see the complete user-defined task list, the overview of today's time spending, and goal progress. You can start/stop time recording simply by tapping a task. It only needs a name and/or icon to set up a unique task, with advanced settings as optional.

Some of the key features of ATracker are
● Ease of use – start and stop an activity with just one tap
● You can view your time history as a list or calendar view.
● Create reports as charts and goal progress, and share via email and social networks.
● Set daily and weekly goals
● Track your time from any device: mobiles, tablets or computers or Apple Watch.

The app is available for free with a one-off payment to remove restrictions or a monthly fee for advanced options.

Available at: atracker.pro

PHYSICAL COGNITIVE VISION DYSLEXIA

Audible

- Software/App (mobile)
- Free trial then a choice of subscriptions £-££ per month or £££ per year
- Education – Social – Leisure

Audible is a subscription-based online audiobook and podcast library.

It works on PC, Mac computers, Amazon, and both Apple and Android devices. Users can stream or download books to their library and listen to them at will, being able to play or pause as required.

Users with a visual impairment can access features such as menus and search functions using methods appropriate to their equipment.

For those accessing Audible through the website, an accessible version of the site is suitable for use with screen readers. For Apple users, the iOS Audible app is supported by Apple VoiceOver.

Audible offers a variety of subscription options. Costs depend on how many books per month the user would like to 'purchase', with options ranging from one per month to 24 per year. However, it is also possible to make a one-time purchase without subscribing.

There is also an accessible version of the Audible website available at www.audible.co.uk

Available at: www.audible.co.uk

VISION HEARING COMMUNICATION PHYSICAL COGNITIVE AUTISM DYSLEXIA

Audio Notetaker 4

- Software
- FREE 30-day trial Annual subscription = £££
- Daily Living – Employment – Education – Social – Leisure

Audio Notetaker 4 is an audio-based software program for easy visual note-taking. It is of particular benefit to users with dyslexia or autism. It empowers users to turn their notes into various formats to suit individual learning styles.

The program is very suited to an educational or work setting. Users can choose to record audio via microphone or import audio from their computer or another device.

It works with Dragon Naturally Speaking Voice Recognition software to transcribe the recording into written text.

Every phrase in a recording is automatically split into manageable sections. Users can then highlight the ones

they think are important/ didn't understand etc, in a choice of colours.

Developer message: "We'll be supporting Audio Notetaker until 2025. But we're now encouraging new and existing users to try Glean."

Available at: sonocent.com

Disabled Black people face multiple disadvantages due to discrimination.

Black Thrive supports this publication because we exist to address the inequalities that negatively impact the mental health and wellbeing of Black people, so that the thriving of all Black communities is the norm.

We want Black communities to be at the forefront of reimagining, redefining and co-creating a society where systemic racism and other oppressive systems have been dismantled and they have fulfilling lives.

To find out more about our work, sign up to mailing list or follow us on Twitter @BlackThrive @BlackThriveLbth and on Instagram @BlackThrive.

Check out our Get involved page for opportunities to connect with us:
lambeth.blackthrive.org/get-involved

lambeth.blackthrive.org

HEARING COMMUNICATION

Ava

   WEBSITE

- Software/App (mobile)
- FREE
- Daily Living – Employment – Education – Social – Leisure

Ava is a live captioning tool that is available as both a software download and an app. A very useful tool for users with a range of hearing impairments, Ava monitors audio and creates closed captions in real-time with 90% accuracy.

Ava for Mac and Windows enables users to view captions along the bottom of their computer screen. Anything from a Zoom work meeting to online classes can be captioned.

The Ava app for Android and iOS makes it easy to live caption conversations on the go. Users simply direct their phone toward the other speaker, and the conversation immediately appears as text on the screen. It is useful for socializing and daily living situations, such as visiting the GP.

Available at: www.ava.me

COMMUNICATION COGNITIVE AUTISM DYSLEXIA

Ayoa Mind Map

- Software/APP (mobile)
- Free 7 day trial/££ per year
- Daily Living – Employment – Education – Social – Leisure

Ayoa Mind Map is a subscription-based software tool/app that encompasses mind mapping, collaboration and task management.

The monthly subscription allows access to mind mapping tools/templates, an image library, file storage and sharing and collaboration systems. There are three types of mind maps available:

- Speed mind maps – allow users to quickly capture ideas in different styles.
- Organic mind maps – inspired by hand-drawn mind maps, use curves to support natural thinking processes.
- Radial maps – are visual pie charts intended to help users prioritise their goals.

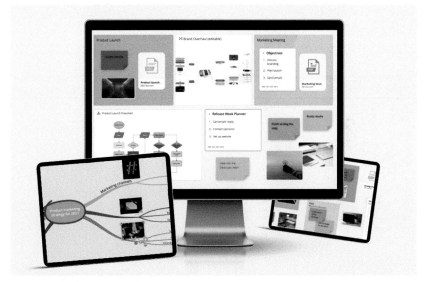

This can prove helpful as visual learning plays an important role in how neurodivergent people process information.

Ayoa Mind Map is the basic app subscription. Also available are the more expensive Ayoa Pro and Ayoa Ultimate, which contain extras such as a Whiteboard feature and built-in video conferencing.

Available at: www.ayoa.com

MENTAL HEALTH

Be A Looper

- App (mobile)
- FREE
- Daily living

The Be A Looper app keeps users 'in the Loop' with up to five chosen contacts or 'Loopers'. These could be relatives, carers, friends or medical professionals.

Each day, users can check how their fellow Loopers are doing and share how their own day is tracking.

Users receive daily prompts reminding them to check-in. The app encourages users to choose a number between 1 and 10 to score how their day is tracking. There is also the option to check in with specific categories such as anxiety, stress and physical health, or customise them.

Once a user has checked in, they can see how their partner Loopers are tracking.

The Be A Looper app is available in 87 countries, so Loopers can check in with each other from across the world.

Please note: At the time of press, contrary to information on the Be A Looper website, the app no longer appears to be available on Android.

Available at: www.bealooper.com

VISION

Be My Eyes

- APP (mobile)
- FREE
- Daily Living

Be My Eyes is an app designed for people with a range of visual impairments and is supported by a worldwide, multilingual community of volunteers.

The app gives users 24-hour access to free visual support

with any task that can be assisted via a live video link. Tasks can be as varied as reading small print, setting up home appliances or choosing clothes that match.

Adult volunteers answer calls from users via a live two-way audio, one-way video link that is opened when a volunteer answers a call. The volunteer can then see what is in front of the user's camera and give verbal support accordingly.

If a task is particularly complicated, there is the option to call one of Be MY Eyes' company partners for Specialised Help via on-demand video customer support. (NB: Specialised Help availability varies by region/company opening hours).

Available at: www.bemyeyes.com

PHYSICAL **COGNITIVE** AUTISM

Beamz

- Software/Hardware/App (mobile)/Content/Consumer Tech/
- FREE/£/££/£££
- Education – Leisure

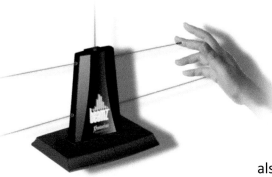

The Beamz Interactive Light Controller Music System enables users to create music simply by touching light beams with their hands. The preset music or sound event will be triggered when the light path is interrupted. It is primarily designed for use in education or therapy settings.

The latest model is the Beamz X4. Its small size means it is easily portable. Its Bluetooth connectivity allows it to be connected to a users' PC or smart device without cables.

Settings can be used to adjust the distance of the light beams. A short space is ideal for individual users, while the extended length allows multiple users to participate at once. There are also options to have the beams always on or only on when the light is interrupted by a hand or object.

The Beamz X4 comes with 150 interactive songs, lessons and therapy activities.

Beamz also produce music-making apps – the Beamz iOS app and interactive virtual music app Jam Studio VR.

Available at: beamzinteractive.com

HEARING

Bellman Maxi Pro Personal Amplifier

- Hardware
- £££
- Daily Living – Employment – Education – Social – Leisure

The Bellman Maxi Pro system is designed to work across various devices to assist users with hearing impairments in sharing in conversations, taking part in phone calls, and enjoying music or watching TV.

The main component in this system is the Bellman Maxi Pro Personal Amplifier which captures and amplifies speech and other sounds. The sound is then transmitted, via a neck loop worn by the user, to their hearing aids, headset or stetoclips.

If a user needs to hear a conversation more clearly, they place the Personal Amplifier close to the speaker.

The Bellman Maxi Pro Personal Amplifier will connect to a user's smartphone and will also boost the phone's ringtone. It uses Bluetooth to link to a tablet, PC or music system to listen to their streaming music. The addition of the Maxi Pro TV streamer will allow users to fully hear television shows.

Its robust and simple to use design is suited to users with low dexterity or hand tremors and those who have some visual impairment.

Available at: bellman.com

VISION HEARING

Bellman Visit

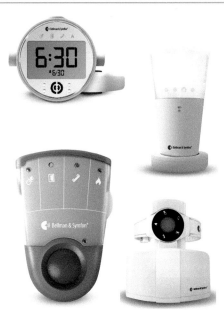

- Hardware
- £££
- Daily living

The Bellman Visit home alerting system is a range of interconnected devices that provide a combination of audible, visual and vibrating alerts.

The system is a useful tool for users with a range of visual or hearing impairments.

Users can choose how they wish to receive alerts, be it flashing lights, loud sounds or gentle vibrations. There is a choice of receivers, pocket pager, wrist receiver, alarm clock (with under the pillow vibration pad) and a flash receiver which can be sited in an easily visible location. Any or all of these can be linked to the system to provide coverage throughout the home.

The Bellman Visit system is easy to set up. Simply plug in the components and switch them on. The stand-alone elements are low power, which can run for years on standard batteries. The components work over a long-range, up to 100ft, so users can pick up alerts even in the garden.

Examples of how the system can be linked include:
- Doorbell rings → Transmitter → Pager
- Phone rings → Mobile Phone Sensor → Flash Receiver
- Smoke → Smoke Detector → Alarm Clock with Bedshaker
- Baby cries → Baby Monitor → Wrist Receiver

Retailers include:
adcohearing.com
www.connevans.co.uk

Info at: bellman.com

VISION PHYSICAL

Big Calculator

- APP (mobile)
- FREE
- Daily Living – Employment – Education

This is a simple and easy-to-see calculator that is suitable for any age.

The app features large Numbers which are easy to press and see. The results of your simple equations are equally well displayed.

The app also can be viewed in landscape mode and offers a Dark UI which can help reduce eye strain.

Available at: apps.apple.com

VISION PHYSICAL COGNITIVE AUTISM DYSLEXIA

Big Launcher

- App (mobile)
- FREE
- Daily Living – Employment – Social – Leisure

BIG Launcher allows you to change the home screen of an Android phone to make it more accessible for people with vision loss or cognitive needs. It offers a simple and easy-to-use interface with the most requested apps available immediately and be customised to your preferences. You can place shortcuts for apps, websites, contacts, etc directly on the home screen. And easily add more screens as needed. It also allows you to hide apps you don't want to use to protect to help find the ones that you most want.

Big launcher also includes an SOS button for use in emergencies

Big launcher also offer the Big Apps Suite which adds a range of popular apps to the phone similar to Big Launcher.

Available at: biglauncher.com

COMMUNICATION

BIGmack

- Hardware
- ££
- Daily Living – Education – Social

This is a simple communication aid for people with visual impairments and those with complex communication needs.

Using the BIGmack you can record speech, music or any other sound directly into the device. You then press the large button on top of the device to have the sound played back.

The BIGmack is an ideal tool for initiating communication and for calling for attention when needed.

Features include two minutes recording time, volume control, appliance jack, interchangeable tops in red, blue, yellow and green.

It is powered by a replaceable 9-volt battery.

Available at: www.inclusive.co.uk • www.ablenetinc.com

PHYSICAL

BIGtrack 2.0 (with switch sockets)

 WEBSITE

- Hardware
- ££
- Daily Living – Employment – Education – Social – Leisure

BIGtrack 2.0 is a USB mouse, incorporating a large 3-in / 7.62-cm trackball for cursor control and two large buttons for left and right mouse clicks. (The colour differentiated buttons are located behind the trackball to avoid accidental clicks.)

The mouse has a chunky design, and its large trackball requires less fine motor control than a standard trackball. This makes it ideal for users who find it challenging to use fingertip control, as it is just as responsive to a swipe with a hand, arm or foot.

BIGtrack 2.0 also includes a Drag Lock feature that enables a user to move objects on the screen without holding the mouse click button down. It is compatible with most laptop, desktop, and tablet computers that can use a standard mouse.

Retailers include:
www.inclusive.co.uk
www.ubuy.com.lb

Info at: www.ablenetinc.com

VISION

Blind-Square

- App (mobile)
- ££
- Daily living

BlindSquare is an accessible GPS app designed for users with a range of visual and hearing impairments.

It works with third-party navigation apps, such as OpenStreetMap and Foursquare to help navigate your way and travel safely, contributing to independence and freedom in daily life.

BlindSquare gathers information about your location, then uses algorithms to determine what information might be of most value. A user can then shake their phone to prompt BlindSquare to announce details about their current address, junctions or crossings in their vicinity and nearby venues.

BlindSquare is available in over 20 languages.

Available at: www.blindsquare.com

VISION

Blindfold Games

- APP (mobile)
- FREE (with in-app puchases)
- Leisure

Blindfold Games declare themselves as being 'dedicated …to bringing the fun to the visually impaired.'

Their flagship game is Blindfold Racer, where players drive with their ears instead of their eyes.

However, they have a catalogue of 35+ games of many types. They have developed audio card games, sports games, casino games, puzzle games, etc.

Blindfold Games have games to suit a range of ages and tastes. Their entire output is audio games, and only a few of them require the player to wear headphones to get the best out of the game.

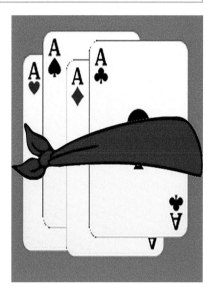

Available at: apps.apple.com

MENTAL HEALTH

Blocksite

 WEBSITE

- Software/App (mobile)
- FREE
- Daily living

BlockSite is an app that aims to help users gain self-control and improve their time management and productivity by allowing you to easily block distracting websites and apps.

The app can be used to schedule tasks and create custom blocklists to prevent the listed websites and apps from opening whilst the user must remain focused.

The BlockSite app can be scheduled to operate between set day hours.

The timer feature also allows a user to set work and rest periods. For example, a 25-minute focused work session followed by a five-minute free browsing period.

As well as user-chosen blocklists – including by category and by keyword – the app also has an in-built 'porn blocker' to restrict websites and other content featuring inappropriate adult content.

You can sync on multiple platforms, blocking the same sits and apps on all your devices. Other features inlcude insights that allow you to explore your browsing trends, and site redirect which automatically redirects when-ever you're tempted to open a blocked site or app.

Available at: blocksite.co

Community TechAid
Strengthening communities through digital empowerment

At Community TechAid our vision is to end the digital divide.

Help us to support our community by donating your old laptop or smartphone!

We wipe, refurbish and recycle unwanted technology helping people online whilst protecting our environment from e-waste

We provide support through organisations in south London, if you're interested in working together or need support please get in touch to find out more

- 🌐 communitytechaid.org.uk
- ✉ contact@communitytechaid.org.uk
- 🐦 @commtechaid
- in @comunitytechaid

Registered charity in England and Wales No. 1193210

VISION PHYSICAL COGNITIVE DYSLEXIA

Bookshare and Read2Go

 WEBSITE

- APP (mobile)/Content
- FREE/££
- Daily Living – Employment – Education – Leisure

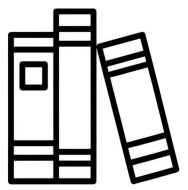

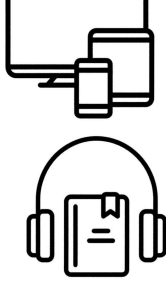

Bookshare is a repository of over a million fully accessible books that is available for anyone who is print-impaired.

This includes many people who are unable to read traditional books as a result of their disability.

This is available in the UK through RNIB Bookshare, which mostly addresses the needs of children and those in education. However, it is also possible for an individual with proof of disability to register on a personal basis with an annual subscription.

Read2Go is an app developed by Bookshare to read accessible books. Whilst still available in the App store it isn't being developed further and other book readers are now recommended.

www.rnib.org.uk/reading-services/rnib-bookshare

Info at: www.bookshare.org/cms

VISION

BrailleBack and KickBack

 WEBSITE

- App (mobile)/Integrated
- FREE
- Daily Living – Employment – Education – Social – Leisure

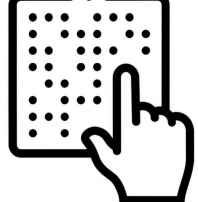

BrailleBack is an accessibility option that helps those who are blind to use braille devices.

It works together with Android's TalkBack app to give combined braille and speech output. You can connect a supported refreshable braille display to your device via Bluetooth.

Screen content is presented on the braille display and you can navigate and interact with your device using the keys on the display. It is possible to input text using the braille keyboard.

The KickBack function adds tactile or haptic signals such as vibrations to your phone to help notify you of messages or other activities that are happening.

Note that the latest updates for Android no longer require BrailleBack to connect to a braille display.

Info at: support.google.com

MENTAL HEALTH COGNITIVE AUTISM DYSLEXIA

Brain in Hand

 WEBSITE

- Software/App (mobile) plus support service
- Monthly Subscription = ££ Cost over a year = £££ Some users may be able to obtain full or partial funding toward costs
- Daily Living – Employment – Education – Social – Leisure

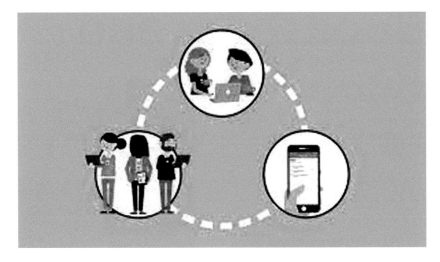

Brain in Hand is a digital self-management support system for people who need help remembering things, making decisions, planning or managing anxiety.

It has a particular benefit to people who have learning difficulties, autism or manage a range of mental health challenges.

Brain in Hand works with service providers to offer round the clock support to vulnerable people. Its aim is to help users live more independently by providing human support combined with digital self-management technology.

The Brain in Hand package consists of three key elements:

- Personal planning sessions with a Specialist to get started and identify a user's strengths and challenges
- An account for secure web and mobile software to access a website where users can plan their day set reminders, break down tasks, add important information, etc.
- Linked support: a connection to responders to help users when they need to get things back on track.

Brain in Hand encourages users to engage in reflection and work with their supporters to review challenges, patterns, and solutions they encounter to help them develop effective strategies.

The Brain in Hand app gives users access to a traffic light system they can use to notify responders when they need support. Responders could be people linked to the user in a professional capacity (e.g., their social worker) but could be relatives or friends. People can only be included in the support network with the user's express permission.

braininhand.co.uk/

Explanatory video: braininhand.co.uk/media/oixexuip/brain-in-hand-system-walkthrough.mp4

The Brain in Hand system 'is approved by government departments and in use throughout health, social care and higher education settings across the UK.'

Available at: braininhand.co.uk

MENTAL HEALTH

Breathe

- App (mobile)
- FREE
- Daily living

The Breathe app for the Apple Watch is designed to help users reduce their stress and improve overall health. It guides users through a series of deep breaths, helping to keep them centered and calm.

The app's guided breathing animation encourages the user to follow the movement of the shape on the screen and inhale as the shape expands and exhale as it shrinks.

Breathe is customisable to suit a user's individual needs and preferences. The breathing rate can be set between four and ten breaths per minute, while the session can last from one minute up to five minutes. Users can opt to receive reminders to start a Breathe session and control those notifications' frequency.

Only available on Apple Watch, but users can track mindful minutes from Breathe sessions in the Health app on their iPhone.

Info at: www.apple.com

PHYSICAL COGNITIVE AUTISM DYSLEXIA

Bubbl.us

WEBSITE

- Software
- FREE/£
- Daily Living – Employment – Education – Social

Bubbl.us is a mind mapping to make connections between ideas or pieces of information. Ideas are linked by lines, creating a diagram of relationships that you can understand at a glance.

This makes it easier to organise your ideas visually in a way that makes sense to you and others.

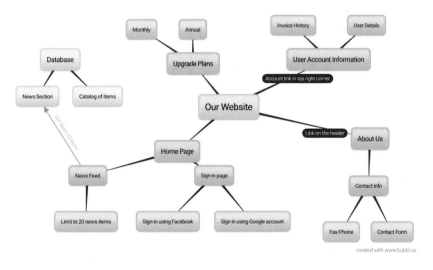

Bubbl.us lets you share your mind maps with people or on social media.

You can collaborate with friends or colleagues and see changes to the mind map immediately.

Because Bubbl.us is browser-based, there is nothing to download, it uses the browser on your phone, tablet or computer.

Available at: bubbl.us

VISION PHYSICAL

Button by Neatebox

- App (mobile)
- FREE
- Daily living

Button by Neatebox is a mobile app that turns the phone into an automatic button presser to operate pedestrian crossings and open powered doors.

The app works with compatible installations with Button equipment installed. In the case of pedestrian crossings, this consists of a special banner being wrapped around the pole just above the control box. When the user stands next to the crossing, the app connects via Bluetooth and automatically presses the button. It works even if the user's phone is still in their bag or pocket.

Button enabled crossings are not yet available everywhere. Via the app's Help section, you can request Neatebox to contact the user's local council on their behalf to discuss installing Button equipment in their area.

Available at: neatebox.com/button

VISION

Cash Reader: Bill Identifier

- App (mobile)
- £
  In-app option to buy Full Version for Lifetime = £
- Daily living

Cash Reader: Bill Identifier is an app that uses a device's camera to identify banknotes from a wide range of global currencies.

It is a useful tool for users with visual impairments – the user simply points the camera at a banknote, and once the app has identified it, it will speak its denomination aloud and show the information on the user's screen in large characters.

Users can download the app for a free 14-day trial, but the app will only recognise the two or three lowest denominations for each currency. Users will have to buy the full version to have full access to all banknote denominations.

Available at: cashreader.app/en

COMMUNICATION

Cboard

 WEBSITE

- Software
- FREE
- Daily Living – Education – Social

Cboard is a free and open-source communication tool that can be used on most devices including desktops, tablets, and mobile phones.

Offline support is available on Google Chrome (desktop & Android). Cboard comes with support for 44 languages and is 100% open source supported by a tic community of developers, speech professionals, parents, and institutions to build the project and software.

Cboard uses open-licensed symbols to build communication grids, you can choose different symbol sets and voices based on your location and preferences.

Cboard can be edited depending on the needs of each user. You can delete, add or rearrange content through the settings section of the app.

Cboard is compatible with a range of access devices including switches when plugged into an Android device.

Available at: www.cboard.io • play.google.com

MENTAL HEALTH

CBT Thought Diary

- APP (mobile)
- Free trial, then annual subscription = ££
- Daily Living

The CBT Thought Diary app is a mood tracker and journal that aims to improve the user's mood by guiding them through the steps to identify, challenge and interpret negative thinking patterns.

The app uses tools that have been found effective in therapies. It encourages the user to record their negative emotions, analyse flaws in their thinking, and re-evaluate their negative thoughts to change them into more balanced ones.

It aims to help users recognise helpful ways to deal with their negative behaviours and emotions.

As well as using CBT Thought Tracker to record thoughts, the app also serves as a daily mood tracker and gratitude journal.

NOTE: Though CBT Thought Diary is a useful tool, users suffering from a mental health disorder should also consult a mental health professional or healthcare provider.

Info at: cbtthoughtdiary.com

VISION

CCTV / video magnifier

  WEBSITE

- Hardware
- £££
- Employment – Education

CCTV (closed circuit television) or video magnifiers provide a low vision aid for a full range of people with visual impairments.

Magnifiers are usually made by a combination of a camera and stand, a viewing screen or monitor, lenses to zoom, and viewing settings such as lighting, colour, and contrast to improve the image. There is often a platform for the document or book to help position it easily before magnifying.

A CCTV magnifier offers a high level of magnification on a large screen supporting independence in many settings. Depending on the video magnifier, they can be used for anything from reading your mail, completing puzzles, browsing books, and writing a letter. Some can be connected to a computer to magnify what is on the computer monitor.

PHYSICAL

Changing Places Toilet Finder

  WEBSITE

- APP (mobile)
- FREE
- Daily Living

Changing Places toilets are not the standard disabled or RADAR toilets, although they share some similarities.

At a spacious 12 square metres, a Changing Place toilet is larger than a standard accessible toilet, which means there is more room for users to be attended to by a carer.

As well as the usual

accessible toilet facilities, it also contains an adult changing bench and a hoist (please note, users need to provide their own sling).

These toilets are designed to be accessible to all users with a disability, including those with profound and/or multiple disabilities that create barriers to using a standard accessible toilet.

There are over 13,000 Changing Places toilets across the UK. The Changing Places Toilet Finder is designed to help users locate them. It is available both as an app and a website.

As well as directions plotted on a map, the user also receives information to locate a toilet within a building. Other information includes opening times, whether the bathroom is unlocked or needs a RADAR key and what kind of hoist is available.

Available at: www.changingplacesmap.org

PHYSICAL

Cherry Compact Keyboard

- Hardware
- ££/£££
- Daily Living – Employment – Education

Cherry make a range of keyboards in different configurations, sizes and with different features. These include various standard, wireless and ergonomic keyboards.

They also make a popular compact keyboard, that is slightly smaller than a laptop keyboard, which is suitable for a wheelchair tray or for one handed users where having the greatest reach across keys woth one hand is helpful.

A matching keyguard is also aavailable for those with tremor.

Available at: www.cherry-world.com

VISION DYSLEXIA

Claro MagX

- APP (mobile)
- FREE
- Daily Living – Employment – Education – Social – Leisure

Claro MagX is a magnifier for use with iPhones and other devices on the ios operating system. It uses your device's camera to enlarge text etc.

MagX applies to everyday life as the user can employ it to enlarge small text in books and newspapers to make them readable.

It can be used to study and photograph small objects and labels, to see menus in the dimly lit restaurant. In short, it can be used to make difficult-to-see things seeable.

The Claro MagX uses a wide range of magnification, high contrast and colour viewing options, making the text easier on your eyes.

Available at: www.clarosoftware.com

DYSLEXIA

Claro Ideas

- Software
- N/A Discontinued as stand-alone – included in ClaroRead
- Employment – Education

ClaroIdeas is an easy-to-use mind mapping and idea capture software. It is designed to help users study, plan, research, outline, and present.

ClaroIdeas gives users the tools to create and edit mind maps using visual components. They can link pictures, audio/video files, research notes, and more to fully capture and express their ideas.

Once created, ClaroIdeas mind maps can be inserted into Microsoft Word documents or PowerPoint presentations. They also work well on interactive whiteboards.

Features include
- Option to add audio notes to individual ideas/nodes on the mind map
- It's easy to make changes to the map connectors (e.g., change colour, thickness, add arrows or labels)
- Option to view a linear outline of the mind map contents.
- If users change the outline or the mind map, both views update instantly.
- Layout easily changed with a click – choose between top-down, left-right, radial, tree and free form

ClaroIdeas is designed to be simple yet functional to be accessible to users of a wide range of ages and abilities.

Available at: www.clarosoftware.com

VISION DYSLEXIA

Claro ScreenRuler

- Software
- £
- Employment – Education

Claro ScreenRuler software allows users to highlight or underline a part of their screen.

It is designed to help users isolate text to make it easier to view and process. The 'ScreenRuler' is a coloured strip that can be positioned over whichever text the user chooses. There is also the option of changing the rest of the screen to a different colour or darkening it out. Users have access to a range of ScreenRuler modes. They can alter the ScreenRuler settings to suit their individual preferences and needs.

This tool could be of

particular benefit to users with dyslexia or a visual impairment, making it challenging to focus on specific text.

Claro ScreenRuler is available for Windows and macOS. It also comes with ClaroView.

Available at: www.clarosoftware.com

VISION COMMUNICATION DYSLEXIA

ClaroRead

- Software
- Individual use: ££
  School or company licence: £££
- Daily Living – Employment – Education – Social – Leisure

Claroread is a software programme providing a simple and easy to use screen reader. The basic ClaroRead will read any online text aloud.

At the same time, Claroread Plus and ClaroRead Pro will also work with scanned paper books and documents. Users with visual impairment or conditions such as dyslexia may find it particularly helpful.

Features include:

- Reads text aloud in high-

quality, human-sounding speech.
- Reads in 30 languages using the user's choice of up to 80 different voices and accents.
- Users can highlight any word, phrase or sentence in a choice of colours.
- Users can change font, text and screen colours to make the text easier to read
- ClaroRead suggests words as you type to help with spelling and writing.
- The user has a choice of dictionaries, including subject-specific or prediction dictionaries.
- The software learns new phonetic predictions as you type or can be trained in a particular subject.
- Speaking spellcheck allows users to select the word they want by hearing the alternatives.
- Provides spelling corrections for problem words or phonetic mistakes
- Portability: ClaroRead can be saved onto a USB to be transferred to any of the user's computers.

ClaroRead Plus and Pro support the following apps:
ClaroPDF
ClaroSpeak
Claro ScanPen
Claro Read Chrome (designed specifically for use with websites).

Available at: www.clarosoftware.com

VISION DYSLEXIA

ClaroView

- Software
- Not available as a stand-alone – included with Claro ScreenReader (£)
- Employment – Education

ClaroView is a software tool designed for users who find it difficult to read black-on-white text on a computer screen.

It settles a coloured overlay over the entire screen. This can benefit users with dyslexia or visual impairment as changing the screen's colour and brightness can make the text easier to read.

ClaroView is fully customisable. Users can choose any colour or level of intensity to suit their individual preferences and needs.

Settings are easily changed to account for different light conditions etc.

An added feature is the option to stop notifications and messages appearing on the screen, so users can work without being disturbed and losing focus.

ClaroView is included in Claro ScreenReader.

Available at: www.clarosoftware.com

PHYSICAL

Clevy 2 Keyboard

- Hardware
- ££
- Daily Living – Employment – Education – Social – Leisure

The Clevy 2 Keyboard is a colourful and robust, child-friendly keyboard designed to help teach writing and keyboard skills to young children and users with a range of special educational needs.

At a generous 2cmx2cm, the keys on this Clevy 2 Keyboard are 30% larger than on a standard keyboard. This, in turn, allows for the characters on them to be up to four times bigger than average. The bigger keys/characters are easier for users to see and, therefore, easier to locate and easier to hit.

Another feature is that the characters are printed in a font similar to that traditionally used to teach handwriting which helps to make letters easy to identify. In addition, the keys are colour coded according to type/use: letters, punctuation, actions.

More information as well as keyguards are available at clevy.com

Available at: www.inclusive.co.uk

COMMUNICATION AUTISM **DYSLEXIA**

Clicker 8

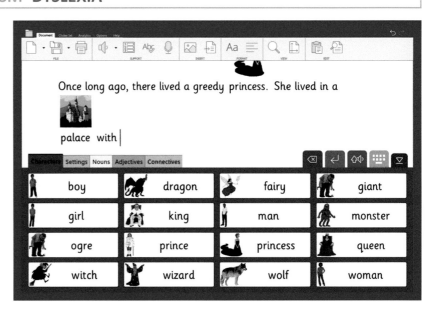

- Software
- £££ (licences for schools)
- Education

Clicker 8 is a piece of software available for licensing in schools. It is a child-friendly word processor aimed at primary level learners.

It helps build writing abilities, starting with simple word recognition and sentence builder grids, and is an effective tool to help students with a wide range of additional needs.

As well as providing support for students with reading and writing difficulties, it can be useful to students who encounter physical challenges with communication. For example, it can be used with switch and eye gaze devices

Free demo available (30 min web session, by appointment): www.cricksoft.com/uk/clicker/free-demonstration

Available at: www.cricksoft.com

COMMUNICATION PHYSICAL **COGNITIVE DYSLEXIA**

Clicker Writer

Chromebook

- Software
- £££
- Education – Social

Clicker Writer (Docs) is a word processor that also reads text aloud at each punctuation mark and offers word choices for misspelled words or from word banks that can be loaded into the app.

It is particularly beneficial for readers who struggle with written expression. The app has word prediction and helps expand vocabulary choices.

Clicker Writer transforms the iPad into a portable primary word processor. Some of the key features include reviewing writing with clear speech and user-friendly lower case keyboard.

The Clicker Docs predictor suggests age-appropriate vocabulary based on the context of your writing. The predictor encourages the use of more adventurous vocabulary, consistently reinforces the correct spellings of the words you want to write, and helps to speed up writing productivity.

There is subject-specific writing support with Word Banks, which provide a speech-supported vocabulary for any subject or topic.

Available at: cricksoft.com

VISION DYSLEXIA

Clover Handheld Electronic Magnifier

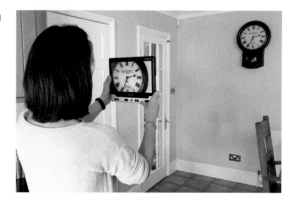

- Hardware
- £££
- Daily Living – Employment – Education – Social – Leisure

Clover Handheld Electronic Magnifiers are battery-operated magnifiers that come in various models. The most recent ones provide HD displays.

In all models, the screen offers a crisp, clear image over a range of magnification levels, making them useful aids for people with visual impairments.

Clover Handheld Electronic Magnifiers come with a range of screen sizes, the largest being 12.5 inches and provide up to 28 x magnification. The smaller ones are easy to hold, but the viewing screens are sufficient to easily access larger materials such as books, magazines and newspapers.

It has an integrated wide-angled reading stand.

Available at: visionaid.co.uk

dasl
disability advice service lambeth

We are a local, user-led organisation run by and for all Disabled people in Lambeth

We are a membership organisation – we are stronger when we work together.

We're keen to hear from you – what matters to you and what you want to change. We are working hard to get your voices heard.

Membership is free - join us!

Our services

- Advocacy Service
- Community Development
- Direct Payments Support
- Specialist Advice Service

Transforming Disabled people's lives

Get in touch

By phone: Leave a message on our enquiry line to request a call back: **020 7738 5656**

By email: enquiry.line@disabilitylambeth.org.uk

Online: Complete the form on our website www.disabilitylambeth.org.uk/contact-us/get-touch

independent living for disabled people

www.disabilitylambeth.org.uk

Charity No. 1087399
A company limited by guarantee registered in England & Wales
No. 04214688

PHYSICAL

Compact keyboards

 WEBSITE

- Hardware
- ££/£££
- Daily Living – Employment – Education – Social – Leisure

As the name suggests, a compact keyboard is smaller than an average keyboard. They are usually around the size of those found on laptops.

They achieve this size reduction not by reducing the size of the keys but by omitting certain keys, usually the numeric keypad.

The advantage of compact keyboards is that their smaller size takes up less space on a desk or lap tray. This is especially useful for users whose individual needs require them to use an assistive control such as a large mouse, joystick, rollerball or other adaptive switches.

The smaller footprint of the compact keyboard leaves room to place such controls next to the keyboard and within easy reach of the user.

A benefit of compact keyboards is that users can position it to suit their needs, rather than needing to place it on a large surface like a desk. It may also bring increased accuracy when using the controls.

Retailers/developers include:
www.cherry.co.uk
www.posturite.co.uk
www.inclusive.co.uk

PHYSICAL

Conceptus Tongue Switch Hardware

- Hardware
- ££
- Leisure

The Conceptus Tongue Switch was developed for use by skydivers wishing to take photos in midair. However, it has since proved useful as a switch for users with physical disabilities.

The Conceptus Tongue Switch has a half-inch pressure-sensitive switch mechanism operated by the user by pressing with your tongue.

Each time the switch is pressed, you feel a click. The switch does not need the user to move their mouth or jaw, as it is completely operated by the user's tongue. This means it is unlikely to fall from the user's mouth.

Available at: conceptusinc.com

VISION HEARING COMMUNICATION PHYSICAL COGNITIVE AUTISM DYSLEXIA

Cosmo Switch

     WEBSITE

- Hardware/Consumer tech
- £££
- Daily Living – Employment – Education – Social – Leisure

The Cosmo Switch is a Bluetooth switch for computers, tablets and mobile phones.

It is highly responsive, making it very accessible for users with a range of motor disabilities.

Its light-up feature makes it a suitable switch for users with visual impairment.

Like many switches, it can be attached to mounting arms and is rechargeable with a long-lasting battery.

It also has many other features that set it apart from other switches.

The Cosmo Switch is highly customisable. Users can adjust settings in areas such as responsiveness, colour, brightness, light mode, battery saving, output character etc.

When pressed, the switch lights up in a colour of the user's choice. This is engaging for the user and provides visual feedback

Other features include:
■ Adjustable activation. Select between ~50 and ~120 grammes.
■ Provides access to switch accessible apps or software across a range of platforms.
■ Access a range of augmented and alternative communication (AAC) such as Jabbla or Tobii-Dynavox.
■ It can be used for switch accessibility, for games or to control music playback.
■ Assign arrow keys and access video games. Easy to set up and easy to pair with the other Cosmo Switches in the same room.

The free Cosmo training app gives access to access to games and activities.

Info at: www.filisia.com/switch

COMMUNICATION   AUTISM

Coughdrop

 WEBSITE

- Software
- £
- Daily Living – Employment – Education – Social

Coughdrop is an AAC (alternative and augmentative communication) app based on open source/open-licensed symbols. You can use it to personalise communication and then use it across multiple devices. You can help grow your vocabulary over time with a simple interface backed up by lots of help and training.

Coughdrop works offline with a backup in the cloud so it is easy to switch devices if something breaks. You can also share boards with others and open-licence them for anyone to use.

CoughDrop is committed to making boards open and reusable. They started the OpenBoards initiative to work toward a standard file format for importing and exporting boards across all apps and devices.

Available at: www.mycoughdrop.com

VISION   DYSLEXIA

C-Pen

- Hardware
- £££
- Daily Living – Employment – Education – Social – Leisure

The C-Pen reader and scanner is a pocket-sized, handheld device that enables users to access and understand text independently.

When used to scan a sentence, the C-Pen reader uses speech synthesis to read it aloud. Or, if the user just wants to know the meaning of a specific word, C-Pen readers have a built-in dictionary and can be used to read out a definition.

The C-Pen comes in a range of models to suit individual needs and circumstances.

■ ReaderPen – a reading pen to support independent reading and learning. Programmed with monolingual dictionaries providing word definitions.

■ ReaderPen Secure – a version of ReaderPen for workplace use.

■ ExamReader – a reading pen for use in exams.

■ LingoPen – for students learning a new language or with English as an Additional Language (EAL).

■ C-PEN Connect – Bluetooth enabled scanner pen/digital highlighter.

Info at cpen.com.

Available at: www.scanningpens.co.uk

VISION

Dancing Dots

- Software
- £££
- Education – Leisure

Dancing Dots offers technology, educational resources, and training to assist people who are blind and have to read, write, and record music.

They seek to promote inclusion, literacy and independence for visually impaired musicians producing and engaging with music.

Lime Lighter helps low vision performers to read magnified print music up to 10 times standard size hands-free.

Those with sight can scan and edit scores and convert them to braille notation with the Braille Music Translator without knowing braille.

Equally, blind musicians can independently create print and braille scores.

Available at: www.dancingdots.com

PHYSICAL

DelanClip

- Hardware/Consumer
- £££
- Daily Living – Employment – Education – Leisure

The Delanclip supports Head Tracking on your PC and uses your head movements instead of a mouse, joystick or another controller. Mostly used as a games controller, especially for simulators Headtracking can reduce or even remove the need to use our hands but can be combined with other technologies such as a switch.

The Delanclip works with your headset and webcam, although they sell recommended items to get the most from the system.

Available at: delanengineering.com

HEARING

Digital Video Baby Monitor Watch

- Hardware
- £££
- Daily living

This wearable Digital Video Baby Monitor Watch enables users with hearing impairments to monitor their child from anywhere in the house.

The Baby Monitor Watch is wirelessly linked to a camera unit in the baby's room. If the baby cries or there is another loud noise.

In that case, the camera unit will trigger a vibration alert on the wrist monitor. The user can view live video on the monitor's screen.

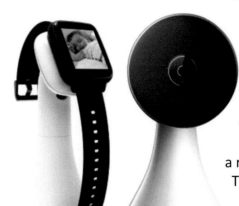

The watch monitor and the receiving unit can form a two-way intercom.

Other features of the system include a remote-controlled setting that will play lullaby melodies from the unit in the baby's room.

The monitor screen can also display the room temperature and a lighting sensor.

When the camera is in night vision mode, it can see over a distance of up to 5m.

The system works over a range of up to 100m. The watch monitor is powered by a fast-charging battery that will last for 6 -10 hours when fully charged. The camera unit can be mounted on a wall or simply placed on a convenient surface.

Available at: easylinkuk.co.uk

PHYSICAL COGNITIVE AUTISM DYSLEXIA

Digital voice recorders

- Hardware
- ££/£££
- Daily Living – Employment – Education – Social

A digital voice recorder is a device that records sound, usually, speech as a digital file that can be easily stored, copied, and played back on a computer, tablet or smartphone With suitable

software the file can be edited even converted into text. Recorders are widely used in employment and education where they can be used to take notes in meetings, record lectures or where researchers want copies of interviews.

Simple recorders are usually handheld with a record button, pause function, and off button. They often have a menu where you can make sure that the file is named and date stamped to make it easy to find. Some only record onto built-in memory, others will record on Micro-SD cards. Many people find it useful to have a recorder with a socket for an external microphone to get better recordings.

Available at: www.dyslexic.com

VISION COGNITIVE DYSLEXIA

Dolphin EasyReader

 WEBSITE

- Software/App (mobile)
- FREE
- Daily Living – Employment – Education – Social – Leisure

EasyReader is a free app designed to make reading more accessible for readers who face challenges in accessing the printed word. It can be especially helpful to users with a visual impairment or a neurodiverse condition such as dyslexia.

EasyReader enables users to choose settings to suit their personal vision and reading preferences. Visual settings include magnifying text or changing colour schemes. Audio settings allow users to synchronise text with speech.

It can also be used with Amazon Kindle Fire tablets and Chromebooks.

Available at: yourdolphin.com

VISION HEARING COMMUNICATION PHYSICAL COGNITIVE AUTISM DYSLEXIA

Doro 1370 mobile phone

- Consumer tech
- ££
- Daily Living – Employment – Education – Social – Leisure

The Doro 1370 is an easy-to-use mobile phone designed for visual and/or hearing impaired users. It has a wide, easy to read display and large, high contrast keys.

These convex, widely spaced keys are suitable for people with poor manual dexterity. Its speakerphone offers loud, clear sounds for clarity when speaking and listening and is also hearing aid compatible. It also features a vibrating ringer and a visual ring notification.

The Doro 1370 has straightforward controls, with three direct keys to take the user to the messaging system, camera or torch. It offers an assistance button that can be pre-programmed with up to 5 numbers. Any phone functions that the user does not wish to access can be hidden.

The Doro's 2.4-inch display screen features large text, with the option to adjust font, colours, and wallpaper size.

This is NOT a smartphone, so it only uses a 2G network. However, as well as being used for calls and texts, it has a 3MP camera for taking photos and video and an MP3 player for music. The microSD card slot allows users to increase the phone's 8MB memory to 32GB for extra storage.

The Doro 1370's battery gives up to 420 minutes of talk time and 530 minutes (over 20 days) on standby.

Available at: www.doro.co

VISION COMMUNICATION PHYSICAL AUTISM **DYSLEXIA**

Dragon Naturally Speaking

- Software
- £££
- Daily Living – Employment – Education – Social – Leisure

Dragon Naturally Speaking is a speech recognition software package. Dragon Home and Dragon Professional are available for use with Windows. At the same time, the Dragon Anywhere mobile app works with any iOS or Android device.

The software's main functions are:
- voice recognition - transcribing speech to written text
- text to speech – speaking aloud the text content of a document.
- recognition of spoken commands

There is a range of functions within these broad features, enabling users to create, format, and edit documents just by speaking. Dragon software is based on a 'Deep Learning' speech engine that continually adjusts to and learns from users' voices.

As well as creating word documents and spreadsheets, users can listen back to what they have written. And there is even the option to use dragon software to turn pre-recorded speech into editable text.

Available at: www.nuance.com

VISION HEARING COMMUNICATION PHYSICAL COGNITIVE AUTISM **DYSLEXIA**

Microsoft Ease of Access Centre

- Software/Integrated
- FREE
- Daily Living – Employment – Education – Social – Leisure

The Ease of Access Centre is part of Micorosft windows settings and is the location where many of the accessibility features can be found. You can open the ease of access center by pressing the Windows key + U.

The software includes:
- magnifier to enlarge all or parts of the display
- high contrast to choose the colour contrast you prefer for the screen
- text or visual alternatives to sounds
- onscreen keyboard
- narrator – hear text read aloud
- keyboard shortcuts to reduce the need to use a mouse

- keyboard shortcuts for touchscreen
- speech recognition.

Similar features can be found on other devices but may be referred to as accessibility options.

Info at: www.microsoft.com/en-us/accessibility

123

HEARING

Echo Bluetooth Transmitter

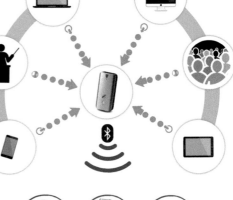

- Hardware/Consumer tech
- £££
- Daily Living – Employment – Education – Social – Leisure

The Echo Universal Bluetooth transmitter will pair with any compatible Bluetooth receiver, such as smartphones, tablets or headphones, to transmit sound up to 10m.

It is a useful tool for users with a hearing impairment, whether they want to listen to music/TV, attend meetings or lectures, or simply share in conversations.

The Echo Universal Bluetooth transmitter has two built-in microphones, one omnidirectional, one uni-directional. Its rechargeable battery lasts for up to 25 hours, or it can be run from mains electricity.

NOTE: The Echo Bluetooth transmitter cannot be paired directly to Blutooth hearing aids.

Available at: www.hear4you.com

VISION   COMMUNICATION   PHYSICAL   COGNITIVE   AUTISM   DYSLEXIA

Emporia SMART.4 bigscreen smartphone

- Hardware/Consumer tech
- £££
- Daily Living – Employment – Education – Social – Leisure

The Emporia SMART.4 is an Android phone with a large 5-inch touchscreen and several features designed to make it accessible to visual or hearing impaired users.

The screen is easy to read, with large text and icons almost one-inch square, and

the phone also has hearing aid compatibility.

It comes pre-programmed with features users may find useful, including an app to magnify text and objects, and a QR scanner. It is equipped with Near Field Communication technology (NFC) which enables the phone to make contactless payments.

A useful feature is an emergency call function. You can pre-set five contacts that the phone will call, message, and notify of the user's location when the emergency button is pressed.

Available at: www.emporiatelecom.co.uk

PHYSICAL

enPathia

 LINUX

- Hardware
- £££
- Daily Living – Employment – Education – Social – Leisure

EnPathia a is an alternative to a computer mouse. It is a movement sensor device most commonly attached to a user's head, so it is often described as a 'head mouse'. However, the device is intended to be adaptable to suit individual users according to their needs. Therefore, the EnPathia come with an adjustable strap that can attach it to any part of the user's body capable of movement. For example, the forearm, wrist or foot. It can also be easily configured to consider different aspects of a user's movement, such as speed, sensitivity and range.

The EnPathia is designed as a tool for users with limited or no mobility in their upper body. For example, users suffering from spinal cord injuries or muscular dystrophy.

Once the sensor is attached, natural body movements allow users to control onscreen any mouse functions such as a cursor, clicking, selecting, dragging and dropping. It is also possible for the user to write by using the cursor to operate an onscreen keyboard.

Although tailored to the individual, EnPathia can be shared with other users. Each user can have their own customised profile.

Available at: www.inclusive.co.uk

VISION HEARING COMMUNICATION **PHYSICAL** **COGNITIVE** **AUTISM** **DYSLEXIA**

EquatIO

 WEBSITE

- Software
- £££
- Daily Living – Employment – Education – Social – Leisure

EquatIO allows you to create equations, formulas, and more, digitally. This is helpful in making maths and STEM more accessible and engaging for you.

You can use EquatIO to have your maths read aloud, and it can help visualise maths problems, create and customise graphs, and help you to visualise and explore a written equation.

EquatIO includes:

● Equation Editor: create your maths and science expressions by typing right into the editor. Use Prediction to insert fractions, exponents, operators, formulas and more.

● Graph Editor: create and customise single or multiple graphs, plot ordered pairs or tables of points, and more, with this tool powered by Desmos graphing calculator.

● Handwriting recognition: handwrite maths expressions using a touchscreen device or mouse pointer.

● Speech input: dictate equations and formulas aloud. EquatIO understands what you're saying, turning your spoken input into written expressions.

● Screenshot reader: turn any equation across the web into accessible, editable maths with the EquatIO Screenshot Reader.

Available at: www.texthelp.com

Assistive Technology Resources for practitioners

The London Office of Technology and Innovation has created resources to help organisations measure the impact and effectiveness of assistive technologies (AT), these include:
- **Case study library**
- **Guide to designing and evaluating pilots**
- **Framework for design and evaluation of AT trials**
- **Templates for user agreements, results**
- **Research on Assistive Technologies, Smart Water Bottles, Amazon Echo**

Visit our website for more info: www.loti.london/projects/assistive-technology

The London Office of Technology and Innovation (LOTI) helps London boroughs work together to bring the best of digital and data innovation to improve public services and outcomes for Londoners.

PHYSICAL

Ergonomic mice

- Hardware
- ££/£££
- Daily Living – Employment – Education

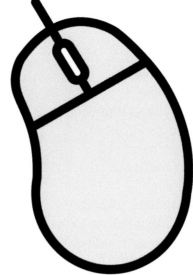

A mouse is used with computers to point and click on icons in Windows or other computer systems.

Most standards mice are designed as 'one size fits all' but an ergonomic mouse is intended to address the health and comfort of the person using it.

Importantly it seeks to minimise discomfort and prevent potential injuries such as RSI, carpal tunnel syndrome and tendonitis.

By being carefully shaped to fit into your hand in a resed position an ergonomic mouse can be useful by helping improve your posture by keeping your hand in a natural handshake position A vertical mouse provides better ergonomic posture and comfort. It is tilted to support a natural, straight forearm posture when being used.

Ergonomic mice also help avoid pressure on your wrist. A standard mouse can place pressure on the tendons and nerves beneath your wrist. This is also reduced by placing your hand at an angle.

Finally, an ergonomic mouse tried to help reduce unwanted muscle activity which reduces repetitive muscle strain. Well-designed ergonomic mice ensure less movement and therefore less fatigue.

As well as ergonomic mice, mouse alternatives such as rollerballs are also effective in reducing pain and strain.

COMMUNICATION

ETran Frame

- Hardware
- ££
- Daily Living – Employment – Education – Social

An E-Tran frame is a sheet of stiff, transparent plastic or (Perspex onto which cards with symbols or words can attached with Blu-Tack or Velcro.

Some people who have communication difficulties also have physical difficulties and find it difficult to point to a book or chart or to handle communication cards.

With an eTran frame the person with a disability points at the words they wish to communicate with their eyes, their communication partner is facing them and can follow the eye movements and confirms which symbols or work they are looking at. The system can work with any preferred symbol set.

Available at: www.liberator.co.uk

COGNITIVE AUTISM DYSLEXIA

Evernote

 WEBSITE

- APP (mobile)/Content
- FREE/£/££/£££
- Daily Living – Employment – Education

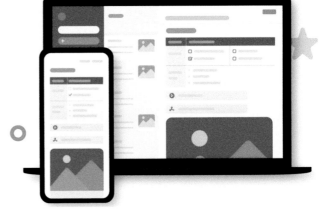

Evernote is an application to help you to keep your, life, work and study organised. It is helpful for note-taking, project planning, and is a way to find what you need when you need it.

The app allows users to create notes, which can be text, drawings, photographs, audio or saved web content. Notes are stored in notebooks and can be tagged, annotated, edited, searched, given attachments, and exported.

Evernote is cross-platform, with a web client as well as applications on Android, iOS, macOS, and Microsoft Windows. It is free to use with monthly usage limits and offers paid plans for expanded or lifted limits.

Two special features of EverNote are tags and the Web Clipper. Tags work like a hashtag. This gives you a different way to organise notes and are useful for searching your notes and categorising them for later use.

The Web Clipper is a browser extension that lets you grab images, text, and even whole web pages. These can be sorted into your Notebooks and you can add tags as you clip.

Available at: evernote.com

PHYSICAL

Evo+

- App (mobile)
- FREE
- Daily living

Evo+ is a useful tool for users who face physical barriers to independent living.

Evo+ is designed to work with Apple technology to turn a user's iPhone or iPad into a universal environmental control device.

The app integrates seamlessly with peripherals around the home. This enables users to remotely carry out everyday tasks, such as unlocking doors, opening curtains, changing TV channels, answering the phone, and raising the alarm for help.

Evo+ is easily operated by both touch screen and input device controls. It allows users to customise their home screen to prioritise their most often used icons and/or change the icon size or colour scheme.

Other features include:
● Audio and visual prompting.
● New keyguard option.
● Fully integrated with the Apple system for automatic updates.
● Updated IP intercom system.
● Utilises wifi, radio and infrared technologies.

Compatible with Steeper and other third party peripherals for easy upgrading of current systems.

Available at: www.steepergroup.com

PHYSICAL

EyeHarp

- Software
- Basic version = Free
  Full version: free for 1 month – then pay subscription or revert to basic version
  Monthly subscription =£
  Yearly subscription or one-off purchase = £££
- Leisure

The Eyeharp is an electronic musical instrument controlled by the user's head movements or eye-gaze technology.

It is designed to enable users with physical disabilities to learn and play music. It has inbuilt eye-tracking hardware, and its specially designed software can define chords and arpeggios or change those definitions and play melodies. The Eyeharp adapts to the individual user's level of ability, suitable for users with a range of musical experiences.

The basic Eyeharp software is available for free or the 'full' subscription version. The free version contains all the necessary settings for scales, harmonizing, etc, but only allows the user to access a piano sound effect. The paid-for version features a collection of 20+ sounds and features such as music learning tools, a selection of playable songs and the option to create your own pieces.

Tobii Eye Tracker is required to use eye control. When installing EyeHarp for the first time, you may be need to instal visual c++ redistributables.

Available at: eyeharp.org

HEARING PHYSICAL

FireAngel Pro Connected Alarms

- Hardware/App (mobile)
- Individual items = ££, but the cost of having several interlinked = £££
  App = free, but comes with in-app purchases
- Daily living

The FireAngel Pro Connected system is a range of domestic alarms connected by wifi and manageable via the FireAngel Connected app and Amazon Alexa.

They are a useful product for users with hearing impairments and those who find it physically challenging to access the alarms for monthly testing.

The alarm system can be made up of a selection of smoke alarms, heat alarms and carbon monoxide alarms.

The Pro Connected Gateway unit is supplied with an ethernet cable to plug into a broadband router. This connection links all the devices and enables push notifications to be sent to the user's mobile or another smart device. The user then has the option to silence the alarm.

Retailers include Amazon, Toolstation and Screwfix.

Info at: www.fireangel.co.uk

COMMUNICATION COGNITIVE AUTISM DYSLEXIA

First Then Visual Schedule

- App (mobile)
- £
- Daily Living – Education

The First Then Visual Schedule app is designed for users with developmental delays, who face barriers to communication or who would benefit from a structured environment.

The app assists a user's parent or carer in providing positive behavioural support.

The app is customisable.

Users can record their own voices, and images to support individual needs can be uploaded from the app's image library, the internet or the user's own photos.

The images are then used to create a visual schedule of daily events. The Checklist feature allows each step to be ticked off when completed. Should an unexpected change be necessary, the schedule can quickly and easily be adapted/updated.

Note: The app works with the iPhone (and iPod). If you want to use it on an iPad, download the app FTVS HD – First Then Visual Schedule HD.

Available at: www.goodkarmaapplications.com

PHYSICAL

Fit Weightlift

- App (mobile)
- £
- Daily living

Fit Weightlift is designed to help users create a workout plan and track their progress. Users can simply add their preferred exercises, weights used, and repetitions to the plan.

Analyze your progress, fully integrated with the Health App

Fit Weightlift will calculate the calories burned, total weight lifted and much more. Exercises can be pre-scheduled to be performed on a given day. The app will record the user's exercise history displayed in easy-to-use graphs. Users can set their own goals and targets and add them to their dashboard for easy viewing.

Features include:
- Active Calorie Burn Calculation.
- Workout Rotation.
- Metric/Imperial unit conversion.

Facility to create Notes about a workout.
- Automatic syncing with Apple Watch.
- iCloud support.
- Handoff – to quickly switch between devices.

Available at: apps.apple.com

PHYSICAL

Fit Wheelchair

- APP (mobile)
- £
- Daily Living – Leisure

Fit Wheelchair is an app to help users track their workouts over their iPhone and/or Apple Watch.

It integrates with a user's Apple Health app to log all their exercise data.

Features include a push counter (it's similar to a step counter), GPS route tracking, and heart rate display.

Fit Wheelchair can display a calendar to show weekly workout patterns when viewed on an iPhone.

The app will also break down each day to show how a workout has affected the rest of a user's daily activity.

Available at: apps.apple.com

PHYSICAL

Fitness Thirty

- Consumer tech
- FREE
- Leisure

Fitness Thirty is a free Alexa skill intended to assist users who want to develop, and stick to, an exercise habit.

It works on the concept that frequent small doses of exercise are easier to maintain.

Fitness Thirty gets its name from the thirty-second blocks of activity it promotes.

A user can simply say which type of exercise they wish to do, and Alexa will call out a workout move to fit that type (e.g. plank, sit-ups, etc). She will then countdown for 25 seconds while the user performs the exercise and then allow for a 5-second rest, thirty seconds in total.

The user can then choose the same or a different category and be given the next move.

There is a choice of four types of exercise, 'abs', 'yoga', 'chest' or 'cardio'.

Users can mix or match types of training and do as many or few as they want, stopping at any time.

Available at: www.amazon.co.uk

PHYSICAL

Fitness trackers

- Consumer tech
- ££/£££
- Daily Living – Leisure

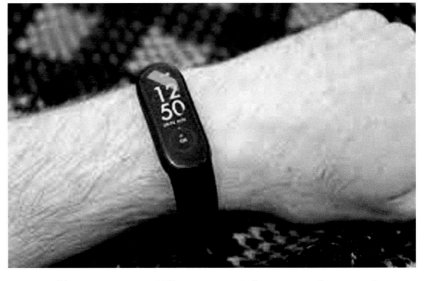

A fitness tracker is a wearable device designed to track and record a user's movements, then convert the data into steps taken, calories burned, sleep quality and information about other activities performed throughout the day.

A tracker can tell whether the wearer is walking, sprinting or even climbing stairs. It can record distance moved and fitness level in terms of heart rate etc. When the user is sleeping, a fitness tracker may register periods of deep sleep and REM so the user can view and analyse their sleep patterns the next day. Most trackers will log the data in a user's smartphone or other smart devices. Users can set goals and view their exercise/ movement history.

A fitness tracker may be a stand-alone device or integrated into another wearable tech such as an Apple Watch. They can be useful to motivate wearers by displaying how close daily goals are or sending tactile, audible, and visual alerts to remind you to increase activity.

PHYSICAL

Flex controller for PC and Nintendo Switch

- Hardware/Consumer tech
- £££
- Leisure

The Flex controller can be used with either switches/joysticks or an eye-gaze controller to play games on a Windows Computer or a Nintendo Switch.

The controller uses the free Controller Settings App for Windows 10 to configure switches or joysticks with orientation and sensitivity settings. It is also possible to configure specific settings for a user's needs, such as 'ignore repeated press period' to counteract the effects of hand tremors. Once programmed, the settings can be saved in profile groups, 6 each for the Nintendo Switch and PC.

The Flex Controller is suitable for any switch connected by a 3.5mm jack plug. Optima joysticks are also compatible.

A Tobii Eye-Tracker 4C or 5 can be used when using a PC.

If the game is being run on a Nintendo Switch, please note that the device must be in its docking station before the Flex Controller can be connected.

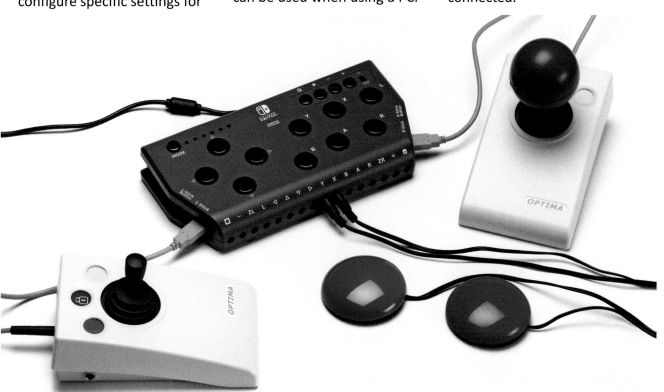

Available at: www.pretorianuk.com/flex-controller

HEARING

fmGenie Radio Aid System

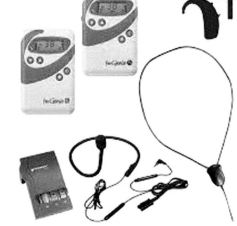

- Hardware/Consumer tech
- £££
- Daily Living – Employment – Education – Social – Leisure

The fmGenie Radio Aid system is a wireless communication system designed to assist users with hearing impairments.

Its main components are a transmitter with a built-in microphone and a receiver. These work in conjunction with the user's personal hearing aids or headphones. Optional extras include attachments to make the system compatible with T-switch hearing aids, cochlear implant processors and bone conductors.

Both the transmitter and the receiver are fully portable. The transmitter is worn by the speaker and the receiver by the user. The transmitter captures sounds and sends them to the receiver, passing them on to the user via their chosen listening aid.

The fmGenie Radio Aid system delivers a clear speech with low distortion levels. The system has an automatic gain control (AGC) which prevents loud sounds (such as the speaker raising their voice) from causing the user discomfort or distorting what they can hear. Because the system delivers sound directly to the user's existing listening aid, it also works to diminish the effect of background sound or other noise distractions.

The fmGenie Radio Aid system allows the speaker to be some distance from the user.

Available at: www.connevans.info

COGNITIVE AUTISM DYSLEXIA

FreeMind

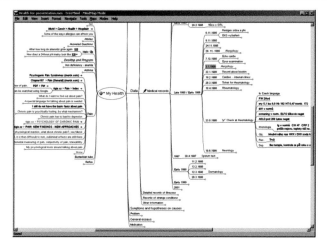

- Software
- FREE
- Daily Living – Employment – Education – Social

FreeMind is free mind-mapping software useful to help users organise their ideas around a central concept.

Features include:
● Fast one-click navigation,
● Undo
● Edit
● Smart Drag and Drop
● Smart copying and pasting into (including pasting of links from HTML)
● Smart copying and pasting from (including both plain text and RTF files)
● Export of map to HTML,
● Find – found items are shown one by one
● Use different colours, fonts and built-in icons
● File mode – users can browse the files on their computer, seeing the folder structure as a mind map.

Available at: freemind.sourceforge.net

VISION **HEARING**

Geemarc Wake 'n' Shake Star Alarm Clock

- Hardware
- ££
- Daily living

The Geemarc Wake 'n' Shake Star Alarm Clock is designed for visual or hearing impairment users.

The clock boasts an extra loud alarm registering up to 95dB at a distance of one metre, while the display consists of 12 extra bright flashing LED lights. Users can also connect a vibrating pad (included) which can be placed under their pillow to provide a physical alert.

The Geemarc Wake 'n' Shake Star Alarm Clock has a programmable alarm that will play for up to one hour. (A snooze option of between 5 and 60 minutes is available).

The display time format is 12/24 hours. The clock also has a lamp function.

The Geemarc Wake 'n' Shake Star Alarm Clock has 5 alarm settings: Off, Vibrator, Ringer + Vibrator, Vibrator + Flash, Ringer + Vibrator + Flash.

Retailers include: Amazon, Hearingdirect.com.

Info at: geemarc.com

COMMUNICATION PHYSICAL **DYSLEXIA**

Ginger

 WEBSITE

- Software/APP (mobile)
- FREE (but there are limits to use)
  Can upgrade to a paid subscription (Monthly = £ but over the year =£££)
- Employment – Education – Social

The Ginger Grammar Checker software gives users access to an AI-powered writing assistant to help them correct and improve their written communication.

The developers claim that Ginger will 'correct 5x more

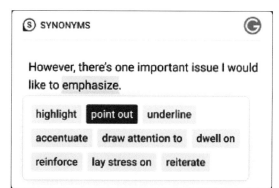

mistakes than Word can detect. 5x faster'. However, as well as correcting simple spelling and grammar errors, the ginger grammar Checker considers the user's whole sentence and offers alternative words or phrases to help refine and develop their style.

Ginger Works across all websites, tools and devices. Whether composing a document in MS Word, sending a business email or simply texting a friend.

Ginger is available for free or as a paid subscription. The free version gives users access up to a certain weekly limit. Once the limit is reached, Ginger will still display corrections for misspelt words. But users have to manually enter the correct word rather than just clicking on the correction.

Ginger is a tool for users who face cognitive challenges in writing. For example, users with dyslexia. But it can also benefit users with a physical condition, such as RSI, by reducing the number of manual corrections and, therefore, the amount of typing they have to do.

Available at: www.gingersoftware.com

SOUTHWARK DIGITAL INCLUSION

FREE INTERNET SUPPORT

**CALL US NOW ON:
07783 776 066**

Southwark Council understands what a vital role technology can play for those with disabilities.

Our digital inclusion team can help show you how to use the internet, complete online forms or seek support with your disability through our network of digital partners.

SOUTHWARK DIGITAL INCLUSION
tel: 07783 776 066
email: digital.inclusion@southwark.gov.uk
web: www.southwark.gov.uk/digital-inclusion

PHYSICAL

Glassouse

 WEBSITE

- Hardware
- £££
- Daily Living – Employment – Education – Social – Leisure

Glassouse is a wearable mouse intended as a tool for users without the use of one or both of their hands or with limited movement. The device looks like a cross between a pair of glasses and a head mouse. It is designed to enable users to control various devices such as computers and smartphones simply by moving their head.

Glassouse is fitted with sensors that allow users to control their mouse through small head movements. The frame housing the sensors is worn by the user and fits similarly to a pair of glasses. A small arm looks like a microphone positioned within easy reach of the user's mouth, but this actually holds the 'bite switch' control. Users move their heads to control the mouse cursor's movement on the screen and use the bite switch to perform the normal mouse click functions of selection, drag and drop, etc.

Glassouse can be used across a range of platforms for various onscreen activities, including gaming. It has wireless connection via Bluetooth and a rechargeable battery that will last for up to 150 hours on a full charge.

Available at: glassouse.com

COGNITIVE AUTISM DYSLEXIA

Glean

  WEBSITE

- APP (mobile)
- Free 30-day trial
  Annual Subscription £££
- Education

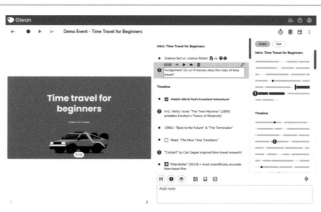

Glean is cloud-based audio note-taking software that helps users work more easily and flexibly. It is primarily intended for an educational setting, although it could also be used at work.

Glean could benefit users with dyslexia or autism facing particular challenges in class, finding it difficult to simultaneously listen, think about meaning, and take relevant notes. Glean captures audio information so the user can review it at their leisure.

Glean is available as a web and mobile app, allowing users to work online and offline, and syncs between devices. There is inbuilt support and training, and free updates.

Glean is based on a four-step process of Capture, Organise, Refine, Integrate (CORI):

- Capture – users record audio in class or online. There is also the facility to screenshot online classes to capture additional information.
- Organise – organisational tools help users to sort, categorise and search multimedia information.
- Refine – users review notes at their own pace to identify and highlight the important parts.
- Integrate – users can mould their notes to create resources as part of their studies.

A useful setting is Reading View, which allows users to remove all distractions, so they can focus on their notes.

Available at: glean.co

PHYSICAL

Glidepoint Touchpad

 WEBSITE

- Hardware
- ££
- Daily Living – Employment – Education – Social – Leisure

Glidepoint is a range of touch and trackpads developed by Cirque.

Glidepoint Touchpads are useful for those with wrist pain or limited wrist mobility, because of conditions such as Repetitive Strain Injury (RSI), Carpal Tunnel Syndrome or arthritis.

The basic model is the Cirque Easy Cat Touchpad which allows basic mouse functions with left and right click activated by tapping on designated areas of the pad.

A subtle change of pad texture and colour helps users identify different areas. Specific taps or glides can execute actions such as scrolling and highlighting or dragging items. The GlideExtend feature makes it easy to keep moving the cursor after running into the touchpad's edge.

Other models include the Cirque Smart Cat Touchpad, which features intelligent software, one-touch scroll and zoom, and distinctive sounds for each operation. The Cirque Smart Cat Pro adds the flexibility of programmable hotlinks to execute programs/commands, control browser functions, assign mouse actions and open files.

Available at: www.cirque.com

COMMUNICATION PHYSICAL **DYSLEXIA**

Global AutoCorrect

  WEBSITE

- Software
- 7-day free trial then download for £££
- Employment – Education – Social

Global AutoCorrect is an intelligent spelling correction software that automatically makes corrections as a user writes. This usually frees the writer from clicking to accept corrections and allows them to focus on their composition without interruptions to their flow and creativity

The software has a vast library of over 130 subject dictionaries covering topics such as law, medicine, finance, design and media studies. Once installed, Global AutoCorrect runs silently in the background.

Global Autocorrect is a useful tool for users who face a range of writing challenges, both cognitive and physical.

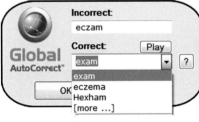 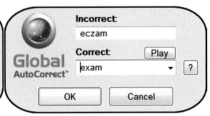

For example, users with dyslexia will benefit from the spelling aspect of auto-corrections, while for users with RSI, the fact that the corrections are automatic reduces the need to correct work manually, thus reducing the amount of typing they have to do.

Available at: www.lexable.com

COMMUNICATION COGNITIVE AUTISM

Global Symbols Board Builder

 WEBSITE

- Software/Website
- FREE
- Daily Living – Employment – Education – Social – Leisure

Global Symbols is a repository of open-licenced symbols for communication that are freely available to all. Symbols are available to support multiple languages, and many are designed to reflect local languages and cultures.

The symbols can be used with Board Builder to create communication boards and grids that can be downloaded and printed as a low-tech resource. They can be exported using an open board format to use other communications tools and devices.

Global symbols also include a symbol creator and editor to allow you to create new symbols or edit existing ones to make them more familiar to you.

Global symbols also offer free and open training materials to help guide you in using the products and symbols and how to implement AAC (alternative and augmentative communication) at home, school or within a community.

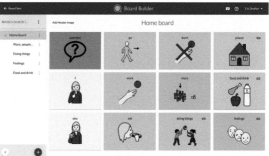

Available at: globalsymbols.com

COMMUNICATION

Go Talk

 WEBSITE

- Hardware
- Go Talk hardware/software = £££
  Go Talk Lite App = Free + in-app purchases
- Daily Living – Employment – Education – Social – Leisure

Go Talk is a battery-powered alternative and augmentative communication (AAC) device and a customisable AAC app.

The hardware version of Go Talk allows a user's contact (a carer or friend, for example)

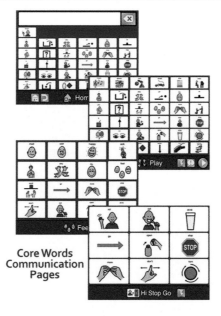

Core Words Communication Pages

to record any message a user may need to use.

The app version works on smart devices and gives access to core words commonly used in daily communication. It includes text-to-speech, adjustable page layouts and customisable navigation.

Go Talk offers users and their assistants the flexibility to import sounds and images from the internet or camera.

The app offers four pages of communication styles: standard, scene, express and keyboard. It has storage for an unlimited number of messages, and pages can be backed up, restored or shared via iTunes/Dropbox.

Both hardware and app versions of Go Talk are useful tools to give a voice to users who face physical or cognitive challenges to speaking.

Available at: www.inclusive.co.uk

COGNITIVE AUTISM DYSLEXIA

Google Keep

 WEBSITE

- Software/App (mobile)
- FREE
- Daily Living – Employment – Education – Social – Leisure

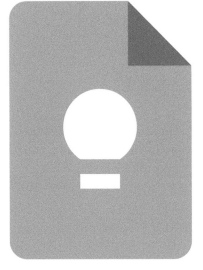

Google Keep allows you to record plain-text notes and organise, edit or share them with others using a suite of collaboration tools.

You can also use Google Keep to create voice notes or set time – and location-based reminders. By using Keep you can quickly capture what's on your mind with a voice memo and have it automatically transcribed. You can grab a photo of a poster, receipt or document and easily organise or find it later when you need it. You can capture a thought or list for yourself, and share it with friends and family.

You can quickly filter and search for notes by colour and other attributes like lists with the label 'to-dos', audio notes with reminders or just see shared notes.

Keep works on your phone, tablet and computer. Everything you add to Keep syncs across your devices so your important files are always with you.

Note that Google Keep is now connected directly to Google Drive.

Information at: www.google.com/keep

VISION

Google Lookout

- App (mobile – integrated or download from Google Play)
- FREE
- Daily Living – Employment – Education – Social – Leisure

Lookout is designed to assist users with visual impairment in exploring their surroundings.

It uses the camera and sensors on an Android device to recognise nearby objects and text and then uses TalkBack to feed information back to the user. It also take pictures and videos.

Lookout operates in five modes: Text, Documents, Food Labels, Currency, and Explore. It also has a Change Language button offering the option of 22 languages (changing a language will also change the default currency).

Text Mode is intended to read short texts such as menus and business cards. For more substantial pieces of text, Document Mode can be used to scan the text and convert it to a digital version, accessible to screen readers.

Food Label mode will not only read the text on a food label but can also be used to scan the item's barcode for information.

Currency Mode cannot identify coins but can identify banknotes, even if they are crumpled or held upside down.

With Explore Mode, the camera reports back what it identifies in the environment, be it text, objects or even animals.

Note: Not to be confused with Android Lookout, which is a personal information security app.

Available at: apps.google.com

VISION PHYSICAL COGNITIVE AUTISM DYSLEXIA

Google Maps

   WEBSITE

- Software/App (mobile)
- FREE
- Daily Living – Employment – Education – Social – Leisure

Google Maps is a free app to help you navigate and find your way in over 220 countries and territories.

Hundreds of millions of businesses and places are included in the maps and you can get real-time navigation, traffic, and transport information. You can also explore local neighbourhoods by knowing where to eat, drink and go.

Features include:

- Help to catch your bus, train or ride-share with real-time transport info
- Group planning made easy
- Offline maps to search and navigate without an internet connection
- Street View and indoor imagery for restaurants, shops, museums, and more
- Indoor maps to quickly find your way inside big places like airports, malls, and stadiums

Google Maps has wheelchair accessibility information for more than 15 million places around the world.

Available at: apps.google.com

COGNITIVE DYSLEXIA

Grammarly

 WEBSITE

- Software
- FREE/££
- Daily Living – Employment – Education – Social – Leisure

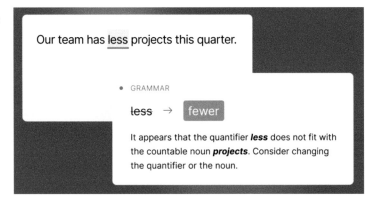

Grammarly is a free to use, cloud-based, cross-platform writing assistant that reviews spelling, punctuation, grammar, clarity etc.

It uses AI to identify errors in the users' writing and search for appropriate replacements. It also allows users to customise their style, tone, and use of context-specific language.

Grammarly is more than a simple spell and grammar checker. In addition to advising spelling and punctuation corrections, it will analyse the user's text for style, tone and meaning and make suggestions accordingly.

Grammarly can be downloaded free from the product website.

Available at: www.grammarly.com

COMMUNICATION PHYSICAL COGNITIVE AUTISM

Grid 3

- Software
- 60-day free trial then £££
- Daily Living – Employment – Education – Social – Leisure

Grid 3 is a software tool designed for users with complex communication or access needs.

It can be used to construct onscreen keyboards, symbol sets, and other AAC (alternative and augmentative communication) aids that can be controlled in various methods.

For example, with eye gaze technology, pointing devices, adaptive switches, voice control or simply by touch.

Grid 3 comes complete with symbol libraries such as Widget, Symbol Stix, Snap Photos and PCS.

Boards can be designed to suit a user's individual needs in terms of symbols and appearance, and control. For example, magnification, cell colour and highlighting, and speed of cursor movement can all be controlled.

Grid 3 also comes with an in-built virtual keyguard to prevent accidental screen presses. Users also have access to audio feedback to give information about cell content, etc.

Available at: thinksmartbox.com

VISION

Handheld magnifiers

- Hardware
- ££/£££
- Daily Living – Employment – Education – Social – Leisure

A handheld video magnifier is a small, lightweight electronic or digital magnifier that is easy to use. and easy to carry with you. You can magnify and increase contrast on the spot to read price labels, maps, bus schedules, directions, and menus.

Handheld electronic video magnifiers allow you to access information wherever you are.

They come in a variety of screen sizes. Larger screens are better suited when you need higher levels of magnification. They usually include rechargeable batteries & can also be used whilst charging.

Magnifiers provide an image on a built-in screen. Most offer a choice of contrast modes and may also save or 'freeze' the image. They may also have the option of attachment to a monitor or television screen.

Even Amazon offers a wide range of magnifiers, including budget and low-cost models.

VISION HEARING COMMUNICATION PHYSICAL COGNITIVE AUTISM DYSLEXIA

HandiCalendar

- APP (mobile)
- Demo version Free for 60 days
  Then you have to buy a licence for ££
- Daily Living – Employment – Education – Social – Leisure

HandiCalendar is a smartphone app to support users in planning and navigating their daily lives. It is a useful tool for any users who need support for personal organisation

HandiCalendar provides users with a visual and audible overview of their day, week, and month. Activities can be assigned images, and alarms

can be set to signal their start and end times. A countdown feature uses a row of dots to indicate time elapsing. There is the option to have text read aloud by a speech synthesizer.

The HandiCalendar app is easy to set up. There is a choice of pre-programmed activities and timers, or users can add their own activities. There is the option to check off completed activities to give the user a visual record of what has or still needs to be done.

To use the handy calendar app, users must first set up an account on myabilia.com. This account can manage the calendar's photos and checklists from any computer.

Users may also choose to allow other people (e.g. their relatives or carers) to access the account. These others can then support the user by monitoring the calendar to see whether the user has carried out their activities as planned.

It can be viewed on Apple Watch.

Info at: www.abilia.com

VISION

HandyReader HD

- Hardware
- £££
- Daily Living – Employment – Education – Social – Leisure

The HandyReader HD is a fully portable, battery-operated, handheld magnifier with easy-to-use tactile controls. A useful tool for users with a visual impairment that makes small text difficult to read.

The 3.5-inch screen offers up to 24x magnification and a freeze-frame function that allows the user to take a

snapshot of the text they wish to read, then zoom in and read it at their leisure.

The Handy Reader HD's small size makes it easy for users to slip into their bag or pocket, so it is always to hand to use with difficult to read text that users encounter in their daily lives; be it the small print on a form, the information on supermarket products or the instructions on a medicine bottle

The HandyReader HD runs on a Replaceable Li-ion battery. Each device comes with a UK power adaptor and charging cable. Also included is an RCA cable that allows users to connect their HandyReader HD to a television, making it easier to view and read longer documents.

Available at: www.visionaid.co.uk

MENTAL HEALTH

Headspace

- APP (mobile)
- 14 day trial = Free, then monthly subscription = £
- Daily Living

Headspace is a meditation and mindfulness app. It offers guided meditations on stress or anxiety management, sleep, personal growth, resilience and mind-body health.

There are hundreds of meditations to choose from. The app promises expert advice to teach users the skills to incorporate meditation into

their daily lives, even if they have never meditated before.

Headspace promises 'Meditation and mindfulness for any mind, mood, goal'. Users are encouraged to begin their day with 'The Wake Up', a daily video series aiming to get the day off to a good start. Then they can choose from focusing on meditations or using 'Move Mode' to access mood-boosting exercise workouts.

At the end of the day, Headspace offers relaxing music, sleep meditations and sleep casts to bring users a restful night.

Whether looking for short mini-meditations on the go, or full courses focused on a particular area, Headspace can provide. Including 'Headspace for Kids', specially designed for younger users.

The Headspace app works with its Chief Music Officer, John Legend, to provide appropriate music to support users' meditations.

Available at: www.headspace.com

PHYSICAL

Helpikeys Programmable Keyboard

 WEBSITE

- Software/Hardware
- £££
- Daily Living – Employment – Education – Social – Leisure

The Helpikeys is an alternative programmable keyboard. It is intended as a replacement for users who find a standard keyboard difficult to use. It is aimed at users who may have limited motor control or dexterity. For example, users with conditions such as Parkinson's.

The keyboard has five pre-programmed layouts and comes with 5 corresponding overlays. A program can be implemented by placing the relevant overlay on the keyboard. (It is held in place by a protective membrane covering the Helpikeys keyboard). In addition, users can design, program and print their own keyboard layouts using the accompanying Layout Builder software.

The five layouts are: Qwerty, Alphabetic, Numeric, Yes/No, Mouse.

The Helpikeys keyboard also includes a programmable five switch interface that can be used for mouse control.

Available at: www.inclusive.co.uk

VISION DYSLEXIA

Hi-vis and high contrast keyboards

- Hardware
- ££/£££
- Daily Living – Employment – Education – Social – Leisure

There is a wide range of high contrast easy-to-see keyboards available. They come in various shapes and sizes and usually have yellow text on a back background or vice versa.

If you have a keyboard that you find comfortable with, you may be able to buy alternative high contrast keys to replace the ones you have. This is usually only possible with older mechanical keyboards and not those which are based upon a membrane.

However, most keys can be made high contrast by buying high contrast stickers which simply stick to your current keyboard to add the extra feature.

Amazon has a range of high contrast keyboards at a variety of prices.

Sight and Sound Technology

Sight and Sound Technology are the UK and Ireland's expert provider of hardware and software for people who are blind and partially sighted and people with learning and reading difficulties.

We recognise that everybody who uses our services, needs specialist technology and support in every part of their daily lives.

We work with individuals, charities, educational organisations, and commercial enterprises to ensure that all users can fully reach their potential, offering not only a large range of solutions, but also the follow-up support necessary to ensure that everyone gets the most out of their technology.

Our products and services that have been specifically designed to improve quality of life at work, at study, or at home.

t: 01604 798070
e: info@sightandsound.co.uk
w: www.sightandsound.co.uk

COGNITIVE AUTISM DYSLEXIA

HotStepper

- App (mobile)
- FREE
- Daily Living – Social – Leisure

HotStepper is an augmented reality navigation app that is easy to use and fun.

The simplicity of 'follow the dancing man' makes it easy for those with autism and cognitive disabilities to follow. The man will lead you to your chosen destination.

HotStepper uses Augmented reality to guide you, this can be especially helpful for those who find maps hard to follow

Available at: hotstepper.dance

MENTAL HEALTH

HourStack

WEBSITE

- Software
- 14 day trial = Free, then monthly subscription = £
- Daily Living

HourStack is a visual time management tool to help plan schedules, track time on tasks, visualise how time is being spent, and make adjustments as necessary.

HourStack users have access to a drag and drop calendar with options for colour coding tasks/projects.

There is a built-in timer, and users can compare actual hours worked on a task with the hours they had scheduled for it. This encourages self-analysis to help users improve their productivity.

Data can be easily exported in a choice of format (Excel, Google Sheets or CSV) and HourStack integrates with various tools such as Google or Outlook Calendars, Todoist, Trello and more.

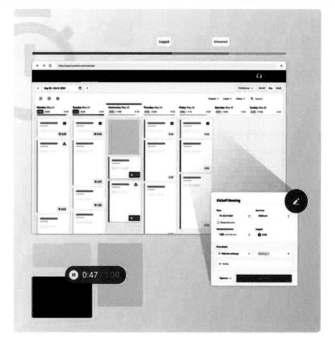

Available at: hourstack.com

PHYSICAL

iCamera Interface Bluetooth for iOS and Android

- Hardware
- FREE
- Leisure

The iCamera Bluetooth for iOS and Android allows users with physical disabilities to use their smartphone or tablet to take photos using an adaptive switch.

It's as simple as plugging the switch into the iCamera interface then pairing the interface to the user's smart device.

Available at: independent-life-technologies.co.uk

PHYSICAL

iClick

- Software/Hardware/App (mobile)/Consumer tech
- £££
- Daily Living – Employment – Education – Social – Leisure

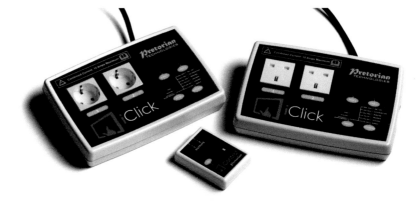

The iClick is a remote control that can be operated via a free app from an iPad. Up to two UK mains sockets can be controlled independently. The iClick also fully supports wireless switches such as the it-Switch.

It is especially useful for people with limited mobility as they can control their environment from one central location. But as it can be used to control many different types of devices, the iClick can also aid communication or cognitive development.

By plugging appliances into the iClick, the user can control them in a variety of ways:

- Direct – an appliance is powered only while the user is touching the switch on the iPad screen.
- Latching – touch once for on and again for off.
- Timed – one touch turns the device on for a predetermined time – choose between 1 second and 250 minutes.
- Cooperative – needs two users to work together to operate the switch.

The app offers four independent and three cooperative control modes. Users who cannot use the touch screen can access the app using the iOS switch control.

The app can display either one or two switches/buttons on the screen. These can be colour customised or replaced with photographs.

The iClick allows interference-free operation over a distance of up to 20 metres.

Note: Download the app via the Pretorian website.

Available at: www.pretorianuk.com

| VISION | PHYSICAL | COGNITIVE | AUTISM | DYSLEXIA |

Immersive Reader

- Software/App (mobile)
- FREE
- Daily Living – Employment – Education

Immersive Reader is a free tool to improve reading and writing for people, regardless of their age or ability.

It can improve reading comprehension and increase fluency for English language learners. It can help build confidence for emerging readers learning to read at higher levels and offer text decoding solutions for students with learning differences such as dyslexia.

It is available in OneNote, Word, and the web version of Outlook. It is also available in Office Lens for iOS.

immersive Reader lets you:
- change font size, text spacing, background colour
- split up words into syllables
- highlight verbs, nouns, adjectives, and sub-clauses
- choose between two fonts shown to help with reading
- read out text aloud, and change the reading speed.

Immersive reader can work with pictures scanned from books. The Optical Character Recognition (OCR) built into OneNote can recognise text within a picture, and then make that available in the app.

Available at: onenote.com

| COMMUNICATION | PHYSICAL | COGNITIVE |

Inclusive MultiSwitch 2

- Hardware
- ££
- Daily living – Employment – Education – Social – Leisure

The Inclusive MultiSwitch 2 is an intelligent switch interface. Simply plug in your switches, and the Inclusive MultiSwitch 2 automatically detects the program you are using and sets up the switches for you.

The Inclusive MultiSwitch 2 can be used with up to six switches at once, allowing multiple users to access a device simultaneously.

Touch-sensitive buttons on the device allow users to

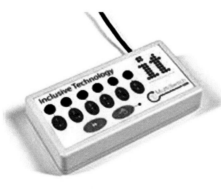

easily test the functions of the switches, or they can use the buttons as switches.

The Inclusive MultiSwitch 2 can also be used to enable switch access to mouse movement. It has four built-in cursor speed settings that are Windows and macOS compatible.

Users can create their own profiles to access non-switch software (e.g. Microsoft PowerPoint or Media Player). They can browse the built-in list and choose mouse clicks, key presses or applications (such as Print or Save). Once selected, they can be saved to the computer. The MultiSwitch will remember the program and automatically set up the switches.

The Inclusive MultiSwitch 2 comes with programmable software. In addition, When online, the MultiSwitch updates itself to work with any new switch software.

Available at: www.inclusive.co.uk

COGNITIVE AUTISM DYSLEXIA

Inspiration

- Software
- ££/£££
- Daily Living – Employment – Education

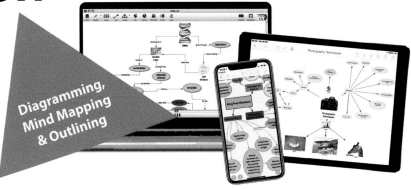

Diagramming, Mind Mapping & Outlining

Inspiration is a simple but powerful visual thinking tool for creating mind maps, concept maps, graphic organisers, outlines, and presentations.

The helps to capture your ideas and visually organise them to communicate concepts and strengthen understanding.

You can transfer your diagram to a written outline which can then be used in other documents and help structure your writing. Inspiration will transfer to Word, PowerPoint, PDF, HTML or as a graphics file.

Inspiration encourages strong critical thinking skills, supports the writing process, and helps you build conceptual understanding to map out your knowledge.

Available at: www.inspiration-at.com

COMMUNICATION PHYSICAL

Integramouse

    WEBSITE

- Hardware
- £££
- Daily Living – Employment – Education – Social – Leisure

The IntegraMouse is a mouth operated joystick that provides hands-free access to mouse functions such as cursor control and activation clicks. It is controlled by the user sipping or puffing through a mouth tube and making slight lip or head movements.

Features include:
- All the functions of a standard mouse device.
- Wireless operation
- Can be used with all standard operating systems that support USB.
- Long battery life with built-in, rechargeable batteries.

- Replaceable mouthpiece system for hygiene
- Built-in bracket so it can easily be attached to a larger stand
- Additional operating modes – joystick or keyboard mode – ideal for gaming.

Available at: www.integramouse.com

| VISION | PHYSICAL | COGNITIVE | AUTISM | DYSLEXIA |

IPEVO Visualisers

- App (mobile)
- £££
- Employment – Education

The range of IPEVO Visualisers allows users to capture, modify and display the video feed from their computer's camera. They work well with document cameras or camera feeds such as iDocCam.

Different models have slightly different features, but all can be used for live presentations and have the same display options. The camera image can be mirrored, rotated, zoomed in/out, snapshotted, have a video filter applied, and have adjustments made to resolution and exposure. In addition, to live presentation, the visualisers can be used to photograph or scan text, photos, QR codes, etc.

IPEVO visualisers can be connected to adaptive switches, such as a foot pedal, to allow hands-free operation. Captured text can be enlarged to be accessible to users with visual impairment.

NOTE: Visualisers for iOS, tvOS, and Android do not support document cameras connected via USB.

Available at: global.ipevo.com

| COGNITIVE | AUTISM | DYSLEXIA |

It's Done!

- APP (mobile)
- £
- Daily Living

It's Done! is an app designed for anybody who may struggle remembering to carry out routine tasks or who are anxious that they may have forgotten.

It's Done! allows users to save a list of tasks and check them off as they are completed. As each task is checked off, the app displays a bold 'tick', vibrates, and makes an audible click sound.

These three acknowledgements help a user create a concrete memory of performing the task. However, if that is not enough to build memory confidence. In that case, a user can easily check the app to see which tasks have been marked as completed.

The app comes with 40 pre-loaded routine tasks, but users can customise their It's Done! lists to reflect their own lives. There is the option to add notes, icons, and photos to individual tasks and rank them according to priority. Tasks can be one-off events or set up as reoccurring. Pop-up notifications can be scheduled for time-sensitive tasks, such as medicine taking.

The lists are easy to manage via the Overview mode. It's Done! can be used to keep a record of past and future tasks.

All task lists can be saved on the user's phone or uploaded to the cloud to be readily available on other devices. There is also the option to notify other people when a task is complete. The app will generate an email or text to send to pre-selected recipients, thus allowing the user's relative, friend or carer to monitor their activities and offer support.

Available at: www.itsdoneapp.com

VISION

JAWS

- Software
- £££
- Daily Living – Employment – Education

JAWS – Job Access With Speech – is a popular screen reader, developed for computer users whose vision loss prevents them from seeing screen content or navigating with a mouse. JAWS provides speech and braille output for most popular computer applications.

It helps to navigate the Internet, write documents, read emails and create presentations.

JAWS can assist by:
- Reading documents, emails, websites and apps.
- Helping navigate with your mous.e
- Scanning and reading documents, including PDF.
- Filling out web forms.
- Providing Easy to use with Daisy formatted basic training.
- Saving time with Skim Reading and Text Analyzer.
- Surfing the net with web browsing keystrokes.

Other features include drivers for all popular braille displays. It works with Microsoft Office, Google Docs, Chrome, Edge, Firefox and much more.

Available at: www.freedomscientific.com

VISION HEARING

Jenile Security Alert System

- Hardware
- £££
- Daily Living

The Jenile Security Alert System is designed for users with a hearing and/or visual impairment. It is a series of sensors that connect via Bluetooth to a centralised alert unit. When a sensor is triggered, the alert unit displays a bright red light to give a visual warning to the user. Users with a visual impairment have the option of a vibrating alert unit that they can place under their pillow.

There are various alert units, including the Light Cube and the Eight-Light Hub. The latter uses its 8 different LED lights to indicate the source of the alert. There is also a vibrating alert unit that can be placed under the user's pillow.

The Jenile Security Alert System sensors include a Smoke and Fire Detector, a Gas Detector, and a Flood Detector. All of which will send a light/vibration alert.

The system also includes a Motion Detector and a Door/Window Detector, which will send a pre-set light or vibration alert to warn of intruders.

Other Jenile Alert products include Doorbells, Baby Monitors and 'bundles' for specific situations, e.g. The Hotel Bundle comprises a Light Cube, a doorbell, an alarm clock and a vibrating pad.

Info at www.jenile.com.

Available at: www.cantoncorecompany.com

PHYSICAL

Joysticks

- Hardware/Consumer Tech
- £/££/£££
- Living – Employment– Education – Social – Leisure

A joystick is a control device consisting of a stick that pivots on a base and reports its angle or direction to the device it is controlling.

Joysticks were often used to control video games, and usually have one or more buttons for extra functions such as 'fire'.

Joysticks are widely used to control other products including cranes, drones, wheelchairs, and surveillance cameras. A range of joysticks can be used to control computers, phones, and tablets as input devices to replace the mouse.

Joysticks come in a range of shapes and sizes, some are designed to sit in a hand and even may be wrapped around the hand and operated with the thumb. When you choose a joystick think carefully about how easy you find it to use, and the equipment or technology you wish to control. Some joysticks are wireless and connect to your device by BueTooth.

A wide range of joysticks suitable for gaming or access can be found on Amazon. Gaming includes Optima Joystick and the Logitech Extreme 3D Pro Joystick.

Joysticks as assistive devices can be found at:

www.pretorianuk.com/joysticks

www.boundlessat.com/Keyboards-Mice/Trackballs-Joysticks

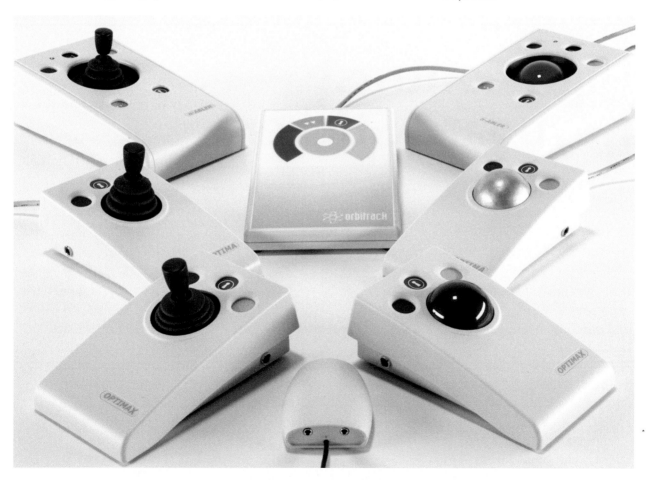

PHYSICAL VISION

Jumbo XL High Visibility Keyboard

    WEBSITE

- Hardware
- ££
- Daily Living – Employment – Education – Social – Leisure

The Jumbo XL II Hi-Visibility Keyboard is designed to have highly visible keys in strongly contrasting yellow and black. This makes it particularly suitable for users with a visual impairment.

Like other keyboards in the Jumbo XL range, this comes with has large one-inch square keys. It has all the standard keys, except for the numeric keypad. All characters are upper case, with black letters on yellow keys. There are separate function keys (f keys) along the top, a Windows key and Shift keys on both sides of the keyboard.

Other features include: Windows or Mac compatible (all keys have a Mac function).

USB connection to a computer

Two extra USB ports on the right-hand side of the keyboard. (suitable for attaching a mouse, trackball, a numeric keypad webcam etc.)

Dimensions: 48.2cm x 17.9cm x 3.4cm.

Available at: www.inclusive.co.uk

VISION DYSLEXIA

High Contrast Keyboard Stickers

- Hardware
- £
- Daily living – Employment – Education – Social – Leisure

The high contrast stickers aim to convert any existing keyboard into a large print keyboard with highly visual keys. They are suitable for laptop and desktop keyboards and are quick and easy to apply.

High contrast keyboard stickers can be ordered online and are usually sold in sets. They are a low-tech aid for people with low vision or dyslexia. Many older people also find these useful.

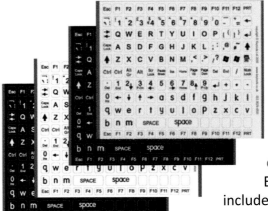

Packs contain sheets of stickers in different colour combinations, designed to suit different needs. The colours are:
- black on yellow
- yellow on black
- black on white
- white on black.

Each set of stickers includes upper and lower case letters, numbers, number pad, cursor keys, symbols, and function keys.

Available at: Adapt-IT.co.uk

PHYSICAL

Keyguards

- Hardware
- £/££
- Daily Living – Employment – Education – Social – Leisure

A keyguard is a plastic or metal plate which fits over a keyboard. They are primarily designed for users with limited motor control who, when using a keyboard, may inadvertently select several keys at once.

When the keyguard is in place, holes in the plate are positioned directly over the keys. This means they can rest their hand on the key guard and make a single key press through a hole without activating any nearby keys. They are also useful for anyone who tires easily when typing. They can rest their hands on the keyguard without pressing the keys underneath.

Most keyboards clip into place, but some are fitted with a lock for extra stability. In any case, they are easily removable.

However, there is a wide variety of keyboard styles and sizes. It is important to buy the correct keyguard to fit a particular keyboard. Most keyguards are designed for a specific keyboard or range of keyboards and therefore are unlikely to fit other keyboards. Because of this, it is common to find keyboards and keyguards sold as a package.

Websites that sell keyguards include:
www.inclusive.co.uk
ceratech.co.uk
www.liberator.co.uk
www.keyboardspecialists.co.uk

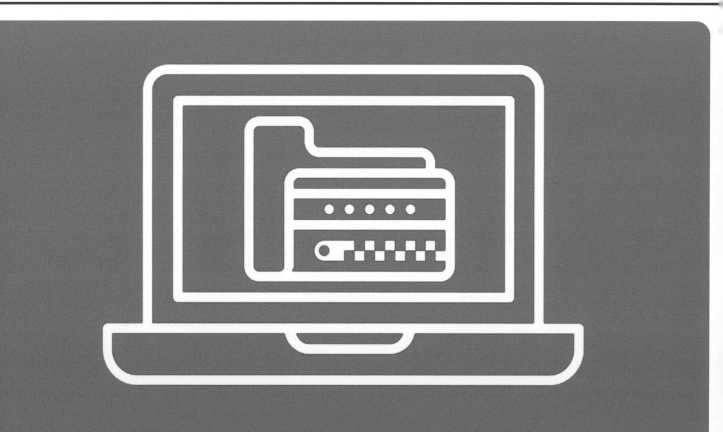

institute of imagination

Join our schools programme

The Institute of Imagination (iOi) is a children's charity providing engaging and interactive educational opportunities for children and young people across arts, digital technologies and science.

Our work champions opportunities for children to develop their imaginations, a quality that is vital to creativity and the next generation's ability to adapt and thrive in a rapidly changing world.

We work with a wide range of schools across our new schools programme.

Find out how you can get involved at www.ioi.london/schools

VISION

Kidde 10SCO Combination Smoke and CO Alarm

- Hardware
- £
- Daily living

The Kidde 10SCO Combination Smoke and CO Alarm is a battery-powered combination alarm featuring both an LED and a voice alert in the event of smoke and/or carbon monoxide detection.

This vocal aspect is particularly beneficial for those with visual impairments. They might be able to hear an alarm but not access visual information (such as a flashing light) that tells them what has triggered the alarm.

The voice alarm not only gives warnings of current dangers, but its Peak Level memory will also notify the user if dangerously high levels of CO have been detected earlier.

The alarm unit features the usual test/reset button.

It has a 'Hush' button to silence accidental alarms or silence the alarm when the danger has been averted. It has a front-loading battery compartment for ease of battery replacement and comes with a 10-year warranty.

The Kidde 10SCO Combination Smoke and CO Alarm gives the following vocal warnings:
- 'FIRE! FIRE!'
- 'WARNING! CARBON MONOXIDE'
- 'Caution, carbon monoxide previously detected.'
- 'Hush Mode Activated'
- 'Hush Mode Cancelled'
- 'Low Battery'

Available at: www.kiddesafeteurope.co.uk

VISION

King's Corner

- App (mobile)
- £
- Leisure

Kings Corner is an app-based solitaire card game.

Players can play against an AI opponent, play with friends online, or even a person using the pass and play mode.

The app is fully compatible with VoiceOver on iOS devices, so it is an accessible game for players with a range of visual impairments.

The developer is Bitpress – www.bitpress.com

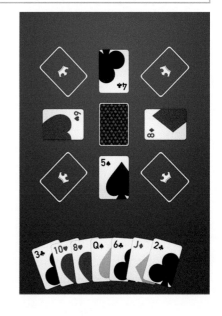

Available at: apps.apple.com

VISION HEARING COMMUNICATION **PHYSICAL** **COGNITIVE** **AUTISM** **DYSLEXIA**

Life 360

- App (mobile)
- FREE/£/££
- Daily Living – Employment – Education – Social – Leisure

Life360 connects you to family and friends to protect and help the people you choose to include.

You can sync your chosen contacts in a private, invite-only circle to see each other's real-time whereabouts and make contact as needed. With your consent, you can see an ongoing timeline of past trips, retrace steps, and see stops made along the way. This can be especially helpful for people who easily become confused and disorientated or who may have lost items when out.

You can create place alerts and be notified as anyone comes and goes. This may help those who cannot easily move around the community, such as those with dementia, from becoming unsafe.

Where necessary, the apps can show a direct path to contacts letting you navigate directly to any circle member just by tapping on their photo on your map. No address is needed.

You can also rely upon 24/7 personal safety by sending a silent alert with your location to contacts in your circle or to an emergency contact.

Life 360 mostly works by finding your location on your phone. Some features may also work using an Apple Watch.

Available at: www.life360.com/intl

MENTAL HEALTH

Lifesum

- APP (mobile)
- Basic app = Free but can upgrade to Premium with in-app purchases = £ per month
- Daily Living

Lifesum is a diet and meal plan app that caters to a user's specific food preferences, lifestyle and health goals.

It offers food, water and exercise tracking and a choice of pre-set and customisable 21-day meal plans of four meals a day. The meal plans and recipes fit into various diet regimes such as vegan, clean eating or high protein.

Lifesum has a huge database of food items along with a barcode scanner to help users access and log a food's nutritional information. It has a library of hundreds of easy-to-cook recipes and a Favourites section where users can keep favourite foods, recipes, meals, and exercises. The app also features a Rating System to encourage users to make healthy choices, and a Habit Tracker to maintain those choices.

Other features include digital shopping lists and in-built shortcuts to quickly record repeat meals.

Lifesum can be synced across other health apps like Fitbit and Apple Health. Users can also control Lifesum via Google Assistant or Siri.

Available at: lifesum.com

COMMUNICATION

Lightwriter

- Hardware
- £££
- Daily Living – Employment – Education – Social – Leisure

With a Lightwriter the person who cannot speak types a message on a keyboard, and it is displayed on two screens, one facing the user and a second facing the communication partners.

Text to speech provides speech output, and some models offer the facility to connect to a printer to provide a hard copy. For people who are unable to use a keyboard, some models offer the option of an onscreen keyboard with switch support using a scanning technique. Word prediction is included to make a significant reduction in the number of keystrokes.

One model of Lightwriter, the SL40, has a built-in mobile phone which provides text messaging and the option of voice telephony with the synthesised speech sent to the person on the other end and the incoming speech message being broadcast through the loudspeaker.

The Lightwriter is especially valuable for people with a good level of literacy skills but who are unable to communicate effectively through speech. It works well in social interactions, at work or when learning, even in noisy, sunny outdoor or poorly lit environments.

Available at: www.abilia.com

PHYSICAL DYSLEXIA

Lightkey (free edition)

 WEBSITE

- Software
- FREE
- Employment – Education – Social – Leisure

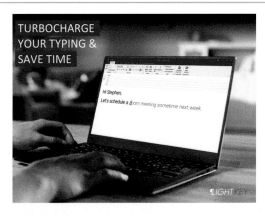

Lightkey is a downloadable, AI-powered writing assistant that predicts and corrects a user's typing as they write.

Lightkey works by learning the user's typing patterns and can eventually predict up to 12 words in advance. To accept a prediction or correct a mistake, the user simply has to press tab and continue typing.

Lightkey offers extra settings specifically designed to improve the usability for people with disabilities:

- The Optimise Prediction setting can be turned on Preferences. It increases the prediction level (over the accuracy level), reducing the number of keystrokes a user will need to make. This can benefit people with disabilities as it minimises the physical aspect of typing.
- The Sound Assistance setting can be accessed from the dashboard. It is designed to help users focus on their keyboard rather than looking at the screen while typing. When Sound Assistance is switched on, Lightkey will play a subtle sound to alert the user to the onscreen suggestion.

Paid-for versions with extra features are also available, such as Lightkey Pro. Lightkey supports Microsoft Office 2010-2021, and Office 365 Chrome and Edge extensions are available to support use on websites.

Available at: www.lightkey.io

HEARING

Live Transcribe

- App (mobile)
- FREE
- Daily Living – Employment – Education – Social – Leisure

Live Transcribe is a smartphone application that transcribes speech into captions in real time. It is available in over 70 languages and dialects. It is a useful tool for users with hearing impairments as they can read what the other person in the conversation is saying.

Live Transcribe uses a smartphone's microphone to capture speech and accurately transcribe it. If one of the users does not want to speak aloud or is unable to do so, a two-way conversation can still take place with the aid of the app's type-back keyboard.

To access Live Transcribe: Go to Settings, Tap Accessibility, Tap Live Transcribe, Tap Open Live Transcribe. To accept permissions, tap OK.

Available at: play.google.com

VISION HEARING PHYSICAL AUTISM **DYSLEXIA**

Livescribe

- Software/Hardware/Consumer Tech
- £££
- Employment – Education

The Livescribe system consists of a digital smartpen digital paper and is supported by software applications.

The Livescribe smartpen looks similar to a standard ballpoint pen, but it has an embedded computer and a digital audio recorder. When used with the Livescribe paper, it records any audio that it hears and uses an in-built infra-red camera to record anything the user writes.

Using the Livescribe desktop software, the saved text can later be uploaded to a computer and synchronised with the audio recording. Users can replay audio directly by simply tapping on their notes or clicking on the relevant portion of the page on the screen.

Handwritten 'pencasts' can be saved as images or pdf files.

LIVESCRIBE+ is the system's partner app that works on all devices.

Users can Sync their Livescribe smartpen to the app and store their notes in the cloud.

Livescribe is a useful tool for users with dyslexia or who face challenges when writing. Difficult to read handwriting can be transcribed into easily read text, supported by the audio recording.

Available at: www.livescribe.com

> PHYSICAL

Logitech Adaptive Gaming Kit

- Hardware
- ££
- Leisure

The Logitech Adaptive Gaming Kit is designed to work with Microsoft's Xbox Adaptive Controller to make gaming more accessible to users with various physical and cognitive disabilities.

The kit consists of a collection of 12 buttons and triggers, two game boards, velcro ties/pads, and coloured labels. The triggers connect to the Adaptive Controller via USB, and the buttons plug into the 3.5mm jacks.

Once the buttons and triggers are connected to the Adaptive Controller, they can be mapped to different controller functions and be used to play games on an Xbox or PC. The buttons can be laid out on the game boards in a layout according to the user's preference for ease of use. They can then be held in place with the velcro ties or pads and labelled to indicate their controller functions.

Set includes 3 large Buttons, 3 small buttons, 4 light touch buttons, 2 variable trigger controls, 2 velcro game boards (one flexible, one rigid). Works with Microsoft Xbox Adaptive Controller.

Available at: www.logitech.com

> VISION

Magnifying Glass + Flashlight

- APP (mobile)
- FREE
- Daily Living

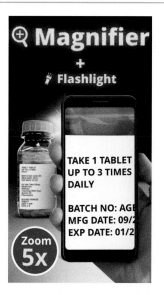

The Magnifying Glass + Flashlight app is, as the name suggests, a digital magnifying tool and led torch combined. It is a useful tool for users with mild visual impairments and can magnify up to 5x.

The app can be used to help the user access print that is too small or too badly lit for them to easily read, e.g. the small print on a medicine bottle or a menu in a dimly lit restaurant. Once activated, users can simply point their phone camera at the text. It will illuminate the subject and engage autofocus, ready to zoom in and out as required. It also has a high contrast mode for easier reading.

There is the option to capture a photo of whatever the user is studying – useful if the text is in a place that makes it difficult to read, e.g. a serial number label on the back of a domestic appliance.

Photos can be stored in the user's photo library.

While primarily designed to magnify print, the Magnifying Glass + Flashlight app can be used to enlarge anything the users require to see more clearly, e.g. threading a needle or removing a splinter.

Available at: apps.apple.com • play.google.com

VISION

Magnifying Glass with Light

- App (mobile)
- FREE
- Daily living

The Magnifying Glass with Light app for iOS turns an iPhone into a lighted, up to 5x digital magnifier.

This app's advantage over simply using an iPhone's camera is that users can normally use the camera or the light, never both at once. Users can simultaneously illuminate the subject of the magnification. This makes it a useful tool for users with mild visual impairment, whether they find the text difficult to read because it is too small or in a dimly lit environment.

Useful features include a high contrast mode, which inverts the colour of the text/background to improve readability. It has autofocus, an image stabiliser, and the option to freeze the image, which can then be moved, dragged or captured as a photo.

The controls are simple and include a finger pinch movement to zoom in or out. A shake of the phone will hide/show the control buttons.

Available at: apps.apple.com

PHYSICAL

Maltron Dual Hand Keyboard

 WEBSITE

- Hardware
- ££/£££
- Daily Living – Employment – Education – Social – Leisure

Maltron is a company that specialises in designing ergonomic keyboards that enable users with various special needs to enter computer data more quickly and easily than with conventional keyboards. Their designs have proven especially beneficial in relieving the symptoms of users who suffer from Repetitive Strain Injury (RSI).

Their fully ergonomic dual handed keyboards come in a range of models intended to fit different shapes of hands and different lengths of fingers to reduce movement and tension. They are available with a standard QWERTY key layout, but purchasers also have the option of the 'Maltron key layout', which is described as being more efficient.

Design features include:

- Two key groups for letters with a central number group enable users to keep their wrists straight and strain-free.
- Letter keys are angled inwards to match natural finger movements.

Maltron offers online adaption training to quickly acquaint users with their 'Maltron pain-free keying system'.

Available at: www.maltron.com

PHYSICAL

Maltron Single Hand Keyboard

 WEBSITE

- Hardware
- ££/£££
- Daily Living – Employment – Education – Social – Leisure

Maltron is a company that specialises in designing ergonomic keyboards that enable users with various special needs to enter computer data more quickly and easily than with conventional keyboards.

Their designs have proven especially beneficial in relieving the symptoms of users who suffer from Repetitive Strain Injury (RSI).

For users who can only type with one hand, Maltron has

developed a single-handed keyboard, available in either a left or a right-handed model.

Both versions are designed for touch typing, and their fully ergonomic shape aids strain and pain-free operationa while still allowing word processing input speeds of up to 85 words per minute.

Features include:
- Carefully planned shape and letter layout, designed to enable the user to access keys quickly and comfortably.
- Keyboard shape matches users' natural hand movements.
- Key arrangement is designed to minimise finger movement.
- Push-On Push-Off keys for Shift, Control and Alt functions.
- Windows keys.
- USB connection to the user's computer.

Maltron offers online adaption training to quickly acquaint users with their 'Maltron pain-free keying system.'

Available at: www.maltron.com

PHYSICAL COGNITIVE AUTISM DYSLEXIA

MathType

 WEBSITE

- Software
- ££
- Employment

MathType is widely used as a way to write maths equations in any digital document. It allows you to write equations with an interface that is friendly and accessible for beginners or studying at an advanced level. MathType adapts to your way of writing maths so you can focus on the task at hand.

If you are using a touch

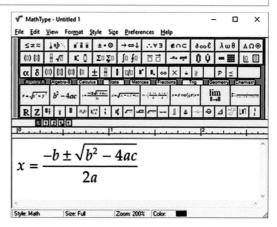

device MathType can convert your handwritten formula into a digital equation ready to be used in your documents.

Available at: www.wiris.com

PHYSICAL

Microsoft Xbox Adaptive Controller

 WEBSITE

- Hardware
- ££
- Leisure

The Microsoft Adaptive Controller works with a range of adaptive switches, joysticks etc., to make games on the Xbox and Windows PCs accessible to players with a range of physical and cognitive disabilities.

The controller features a direction pad, menu buttons and two large pressure pads suitable for players with limited fine manual dexterity.

In addition to these controls, sockets (one USB and more than twelve 3.5 mm ports) can be used to connect the external switches, etc.

Once connected, these input devices are mapped to the various control functions necessary to play the user's chosen games.

Available at: www.microsoft.com

COGNITIVE AUTISM DYSLEXIA

Microsoft Editor

- Software
- FREE
- Employment – Education – Social

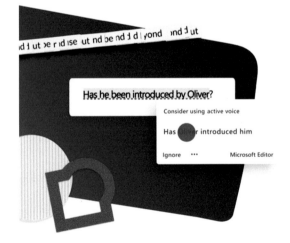

Microsoft Editor is an AI-powered service that helps bring out your best writer in more than 20 languages, whether you are writing a Word doc, composing an email message or posting on a website like LinkedIn or Facebook.

Editor underlines any issues with your writing. You can select any underlined word or phrase and accept or ignore the suggestion. These include
- Spelling suggestions (red underline squiggles)
- Grammar suggestions (a double blue underline)
- Refinement suggestions (a dashed purple underline)

Editor will also check your document for similarity to online sources to help avoid being accused of copying others' work

The basic Spelling and Grammar features are available to anyone with a Microsoft account. This gives you the features to check your writing with suggested revisions and spellings to help.

Available at: www.microsoft.com

VISION HEARING COMMUNICATION PHYSICAL **COGNITIVE AUTISM DYSLEXIA**

Microsoft Office including OneNote

- Software
- ££
- Daily Living – Employment – Education – Social – Leisure

Microsoft Office is the most widely used productivity package in the world. It contains all of the main tools that you would need to produce documents, presentations, connect to people and collaborate.

There are some features that are specific to Office 365 for Business but the core apps and services available for personal or home users include:

● Word – a powerful word processing package for writing

● Excel – a spreadsheet package for working with figures and data

● PowerPoint – for creating presentations

● Outlook – for sending email

● Skype – for voice and video calls over the internet

● Microsoft Teams – for collaborating at work or when learning

● OneDrive – cloud storage for your files.

Office 365 also includes OneNote. This helps you to organise your information and ideas into notebooks that you can divide into sections and pages.

With easy navigation and search, you'll always find your notes right where you wanted them to be.

You can revise your notes from any device with access to your account. You can type, highlight or use ink annotations.

You can use OneNote to

● sort content across notebooks, sections, and pages

● highlight notes with tags

● draw your thoughts and annotate your notes, using a stylus or your finger

● record audio notes, insert online videos and add files

● use the OneNote Web Clipper to save content with a single click

● share notebooks with coworkers, friends, and family.

Office 365
www.microsoft.com/en-us/microsoft-365

OneNote
www.microsoft.com/en-us/microsoft-365/onenote/digital-note-taking-app

Available at: www.microsoft.com

Elim House
Outreach Services
Elderly People

A New community initiative

ARE YOU CONNECTED TO ANY SERVICES? NO

ARE YOU RECEIVING HOMECARE SUPPORT? NO

ARE YOU FEELING LEFT OUT AND ALONE? YES

Pop into Elim House and get support along the way

86-88 Bellenden Road, Peckham, London SE15 4RQ
0207 358 9502 – elimhousecas@gmail.com
elimhousedaycentre.com

VISION

Microsoft Soundscape

- App (mobile)
- FREE
- Daily living

The Microsoft Soundscape app with spatial audio provides the visually impaired user with an audio description of their surroundings. It is currently only available for iPhone and iPad.

It places audio cues and labels in 3D space rather than simply describing nearby objects, hence it sounds to the user as if they are coming from the relevant direction in their surroundings. To access this aspect fully, the user needs to wear stereo headphones or Apple AirPods.

As the user walks around, the app can be set to announce important features as they pass such as roads, junctions, parks, points of interes. There is also the option to place an audio beacon somewhere the user wishes to navigate. 'Nearby Markers' can describe nearby places the user has marked.

Users can orient themselves by using 'My Location' (which describes a user's current location and the direction they are facing) or 'Around Me' (which details the main points of interest to be found in each of the compass directions). Then, as they walk along, the user can use 'Ahead of Me' to describe different features as they are approached.

Available at: www.microsoft.com

COGNITIVE AUTISM DYSLEXIA

Microsoft To Do

 WEBSITE

- Software
- FREE
- Daily Living – Employment – Education – Social

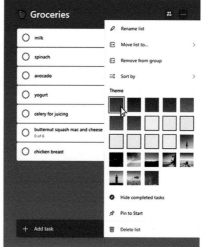

Microsoft To-Do helps you to focus and organise yourself whether that be for work, education at home or when planning social events. It is designed to be a smart daily planner.

To-Do learns from your activity and offers personalised suggestions to update your daily or weekly to-do list. You can manage your to-do list online and it works across platforms whether you're at home using a laptop or travelling with the mobile app you can access and view your task list, edit it to help stay organised.

To-do helps you to break tasks down into simple steps, add due dates, and set reminders for your daily checklist to keep you on track.

Available at: www.microsoft.com

167

AUTISM DYSLEXIA

MindView AT

   WEBSITE

- Software
- Monthly subscription = ££
  Annually -> £££
- Employment – Education

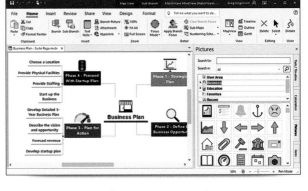

MindView is mind mapping software designed to help users visually organise and manage their ideas, notes and projects.

MindView AT is a version of the Mindview software with specially added assistive technology functionality. It is primarily intended for use in an educational setting (although it could also be useful in an employment setting). MindView AT is designed to help users who prefer to think visually, to produce written assignments and presentations.

It can assist with taking notes during lectures and planning out work. The built-in citation and referencing tool help with the production of academic papers and presentations, which can easily be exported to Word and PowerPoint.

Features include:
● Dragon Naturally Speaking integration with Native Commands
● All Text Made Available for Screen Readers
● All Training Videos Sub-Titled
● Keyboard Only Accessibility for all Functionality
● High-contrast Mode
● Improved Integration with Jaws and ZoomText
● Timeline and Gantt view for Project Planning

Mindview was accredited by the Digital Accessibility Center for individuals with disabilities, including autism.

Available at: www.matchware.com

PHYSICAL

Mouse4all

- APP (mobile)
- FREE – users will need to supply their own switches to use the app with a device
- Daily Living – Employment – Education – Social – Leisure

Mouse4all is an app designed to enable users to use an Android smartphone or tablet without needing to touch the screen. Users can access the internet, games, apps, and social media. It is suitable for users with a physical disability, making it challenging to use a touchscreen, such as users with Parkinson's, multiple sclerosis or cerebral palsy.

There are two Mouse4all systems:

● Mouse4all Switch – use the app with up to two switches. These can be connected with a cable or wirelessly via Bluetooth.
● Mouse4all Box – use the app with a connection box. Users can connect a combination of two of the following: switch/trackball/joystick.

The switch is used to control the app's onscreen cursor – most types of switches are suitable.

Compatible switches and adapters: mouse4all.com/en/compatibility-table

Video Demonstration: youtu.be/HE7I_KoPPG4

Available at: mouse4all.com

PHYSICAL

MyFitnessPal

 WEBSITE

- App (mobile)
- FREE or premium subscription = £ per month
- Daily living

MyFitnessPal is designed for users to track and log calories consumed and exercise calories burned.

It encourages you to keep a food diary that automatically analyses the foods you log to display your nutritional values. Users can also log any exercise carried out. The calories used will be deducted from calories eaten to give a daily total.

MyFitnessPal has a range of features to make it easier to complete the daily diary. A database of basic and processed foods lets you select and input foods according to category/brand, how it has been cooked or prepared and how much it weighs.

If a food is not in the database, you can submit the details for it to be added. Alternatively, the app provides a barcode scanner to check the information on packaging. There is also the option to input recipe ingredients.

MyFitnessPal will calculate the nutritional values of the finished dish, either as an absolute total or per portion.

You can set weight loss and other goals on the dashboard, and the app will give visual feedback as to progress.

A MyFitnessPal account automatically syncs between the user's smart devices, other health apps and the website to provide tracking and information.

Available at: www.myfitnesspal.com

VISION

My Vision Helper

- App (mobile)
- ££
- Daily Living – Employment – Education – Social – Leisure

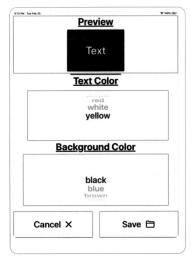

My Vision Helper offers 'advanced camera video magnification, colour contrast enhancement, and Optical Character Recognition (OCR) capabilities.'

The app integrates with Apple's speech recognition software which allows it to be operated almost exclusively by voice commands. This aspect is particularly useful for users with a range of visual impairments. Speech-operated controls include contrast, saturation; magnification; flashlight brightness; rotation locking, and the ability to save load and edit custom filters and colours.

Users also have access to a reference guide and easy-to-follow video tutorials within the app.

In addition, My Vision Helper offers an optional OCR subscription that accurately converts text on the camera to speech.

Available at: apps.apple.com

PHYSICAL

MyPlate Calorie Counter

- App (mobile)
- Free or Gold Membership = £ per month
- Daily living

The MyPlate calorie Counter app boasts a large food database of over 2 million items to help users log and review the foods they consume. It also provides a barcode scanner to easily input information from food packaging.

Users can set their own weight loss and nutritional goals or use personalised

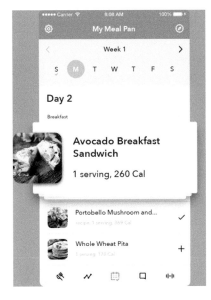

goals suggested by the app. MyPlate will provide a nutritional breakdown for every food eaten so that users can track their daily intake of calories, protein, carbs, fat, sugar etc.

Other features include pre-set meal plans or the ability to input custom food and meals. Users can set up meal-time reminders to keep them on track and record their daily water intake.

The MyPlate app features an exercise database for users to select and input any exercise they do. Additionally, MyPlate will automatically sync across a user's smart devices and other health apps to take account of data such as step count or calories burned through exercise.

Livestrong.com provides a free companion tool to MyPlate, with additional features such as meal plans, workout videos, exercise ideas and more.

Available at: www.livestrong.com

PHYSICAL

mySugr

- APP (mobile)
- FREE or PRO subscription = £ per month
- Daily Living

The mySugr diabetes app is a free diabetes logbook that keeps your diabetes data under control.

The app auto-logs your data, and you can view your daily

information such as meals, your diet, and your carb intake. It records the medication you take and your blood glucose and insulin levels.

Features include:
- personalised dashboard (diet, meds, carb intake, blood glucose levels and more)
- Insulin/Bolus calculator with precise insulin dose recommendations
- blood sugar level graphs.
- estimated HbA1c at a glance, no more surprises
- daily, weekly and monthly reports can be shared directly with your doctor
- Secure data backup (built with regulatory compliance, quality, and safety).

Available at: www.mysugr.com

VISION HEARING COMMUNICATION PHYSICAL COGNITIVE AUTISM DYSLEXIA

NHS App

 WEBSITE

- App (mobile)
- FREE
- Daily living

The NHS app is a one-stop shop to access your health information and services on your phone or through the NHS website.

The app is owned and run by the NHS, and offers a simple and secure online way to:
- get your NHS covid pass
- get advice about coronavirus
- order repeat prescriptions
- book appointments
- get health advice
- view your health record
- register your organ donation decision
- view your NHS number.

To have an NHS account, you must be aged 13 or over and registered wth a GP surgery in England or the Isle of Man.

Available at: www.nhs.uk/nhs-app

VISION HEARING COMMUNICATION PHYSICAL COGNITIVE AUTISM DYSLEXIA

NHS Weight Loss Plan

  WEBSITE

- APP (mobile)/Content
- FREE
- Daily living

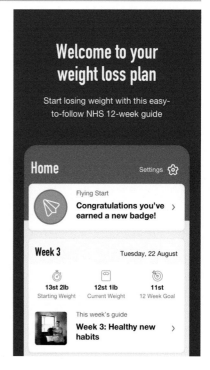

The free NHS Weight Loss Plan is designed to help you start healthier eating habits, be more active and start losing weight.

The plan is broken down into 12 weeks so you can:
- set weight loss goals
- use the BMI calculator to customise your plan
- plan your meals
- make healthier food choices
- get more active and burn more calories
- record your activity and progress.

There is lots of help and support available on the site, including EFL Trust Fit Fans – a free programme for football fans – and guidance on finding out about other weight management and healthy lifestyle services provided by the NHS and local councils.

There is also further information on Type 2 Diabetes and a wide range of low-cost weight loss plans offered through NHS partners.

Available at: www.nhs.uk/better-health/lose-weight

PHYSICAL

Nintendo Wii Switch Enabled Classic Arcade Controller

- Hardware/Consumer
- FREE
- Leisure

The Nintendo Wii Switch Enabled Classic Arcade Controller can be used with any games that support the classic Wii controller to make them more accessible to players with a disability.

All eight action buttons have been adapted to use with an adaptive switch or other control connected by a 3.5mm jack. Adaptive controls could include buttons, joysticks, sip and puff switches, foot pedals, etc.

The Arcade Controllers' in-built joystick has a rubber coating to increase friction. The users can control it by resting their hands on it rather than needing to grip it.

Available at: inclusiveinc.org

VISION

NVDA – Non-Visual Desktop Assistant

- Software
- FREE
- Daily Living – Employment – Education – Social – Leisure

The Non-Visual Desktop Assistant (NVDA) is a free-to-download screen reader that enables people with a range of visual impairments to access and interact with the Windows operating system and other third-party applications such as Microsoft Office, search engines, and music players.

Features include:
- an easy-to-use talking installer.
- a built-in speech synthesizer supporting over 50 languages
- support for many braille displays, including inputs via braille keyboards
- reporting of textual formatting, such as font information and spelling errors
- automatic announcing of all text under the mouse and optional audible indications of the mouse position
- announcing controls and text whilst interacting with gestures on touch screens.

It is 100% portable, as it can be run entirely from a USB flash drive or other portable media without needing to be installed.

Available at: www.nvaccess.org

VISION HEARING COMMUNICATION PHYSICAL COGNITIVE AUTISM DYSLEXIA

OmniPage OCR

- Software
- £££
- Daily Living – Employment – Education – Social – Leisure

OmniPage is an optical character recognition (OCR) app that digitises documents so they can be read onscreen.

This also means that they can be searched and edited. It also has a Photo Conversion feature that converts a picture or photo into a readable format, can be accessed on mobile devices and electronic book readers. It is available in a range of packages, the most basic one being OmniPage Standard which is outlined here.

OmniPage exhibits a high standard of word accuracy and maintains a document's original formatting. Hence, the onscreen version looks the same as the original. The operation is user-intuitive, with easy-to-use tools and flexible options. Users can take advantage of the built-in one-click workflows or create their own.

OmniPage supports a range of output formats, including Microsoft Office, PDF, and Amazon Kindle. It also supports more than 120 different languages.

Available at: www.kofax.com

VISION COMMUNICATION PHYSICAL **COGNITIVE** AUTISM DYSLEXIA

Onscreen keyboards

- Software/Integrated
- FREE/£/££/£££
- Daily Living – Employment – Education – Social – Leisure

Onscreen keyboards are a standard feature of smartphones and tablets. They are a software feature that allows users to type on a touchscreen without a physical keyboard.

They are also available as an option for desktop computers. For example, an onscreen keyboard can be accessed through the settings in Windows 10.

An onscreen keyboard may simply recreate the layout of a standard physical keyboard or vary in terms of which characters are available and how they are laid out. This could be due to the size of the screen.

For example, a mobile phone keyboard would normally need to alternate between two screens, one for letters and one for numbers and punctuation. Or it may be dependent on the requirements of the programme being used.

An onscreen keyboard may

be an accessible alternative for users with a physical barrier to using a traditional physical keyboard.

There is no need to press the keys – a simple touch is enough. Some onscreen keyboards allow users to drag their fingers across the screen to trace a word or command.

Alternatively, an adaptive mouse and/or switch can be used to activate the keys.

Online keyboards can also be controlled with eye-tracking technology. Onscreen keyboards designed for users with a specific disability may be made more accessible in terms of key size, colours, and contrast.

In addition to the free integrated keyboards, there is also a range of commercial onscreen keyboards that might add extra features that can be useful.

PHYSICAL

One Switch Games

- Software
- FREE
- Leisure

One Switch Games are designed to be accessible via a range of adaptive switches, such as buttons, pressure mats, sound switches, movement switches, and many more. The One Switch Games website has a Help section with easy-to-follow instructions on how to download any of their games. Once a game is downloaded, the player's switch or switches should be assigned to particular keystrokes necessary to play the game. (The player's own adaptive switch interface should come with instructions on how to do that.) Most One Switch Games are operated by pressing the Space Bar, so the switch can be mapped to that.

One Switch Games offer games over a wide range of categories. For example, puzzles, race games, cause and effect games, shoot-em-ups, arcade games, etc.

Available at: www.oneswitch.org.uk

VISION

OneStep Reader

- Software/App (mobile)
- ££
- Daily Living – Employment – Education – Leisure

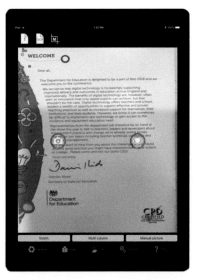

OneStep Reader is designed for users with a range of visual impairments. The basic premise is that the user takes a photo on their smartphone of any printed text they find difficult to read. OneStep Reader will then read the text aloud in clear synthetic speech. The user can access it through a connected braille display.

The OneStep Reader app is easy to use. Its Viewfinder assist, tilt assist, and automatic text detection features ensure that users photograph the whole page of text. User-selectable options include background colour, double highlighting, font colour, size and type.

The OneStep Reader app also provides a synchronised text highlighting feature, which reads along with the displayed text. This is a useful feature for users with dyslexia or similar barriers to reading. The app allows navigation by a single word, sentence or line.

The OneStep Reader is a useful tool to access a range of texts that a user might encounter in their daily lives. It offers different modes for labels, price tags, books, articles, invoices, etc. It can read documents with multiple pages and export to the cloud as HTML or TXT files.

Previously known as KNFB Reader.

Available at: knfbreader.com + www.onestepreader.com

GLOBAL SYMBOLS

A place to discover and publish Alternative and Augmentative Communication (AAC) symbol sets

Many people use symbols as a communication aid to breakdown language barriers and they can also provide support for those with complex communication needs. AAC "is a range of strategies and tools to help people who struggle with speech. These may be simple letter or picture boards or sophisticated computer-based systems. AAC helps someone to communicate as effectively as possible, in as many situations as possible." (Communication Matters – what is AAC?)

For Symbol Users, Friends, Family and Carers

Global Symbols has a huge collection of high-quality symbols you can use for communication. Search our symbols, download and use them. With our free Board Builder and Symbol Creator, you can quickly create boards of symbols to help communication, for day-to-day activities, special occasions, special topics such as games or mealtimes etc. The templates are easy to use for making information sheets or any curtomised designs you want to create.

For Symbol Set Creators

Get your symbols online quickly, where they can be discovered and used by anyone. Global Symbols is a free publishing platform designed especially for symbol sets. You'll spend less time managing a complex website, so you can concentrate more on your symbols and your community.

- Visitors can browse and search your symbols in a rich catalogue.
- Multi-language symbol labels
- Upload draft symbols, publish later
- Review and voting tools to get feedback from your community
- An API for your symbol set
- Automatic inclusion in Board Builder and several other online symbol tools

Our Project

Global Symbols aims to link and create freely available AAC symbol sets with other linguistically and culturally localised symbol sets to provide world wide access to appropriate pictographic based communication that can be used on any communication application. We have worked with several app developers who were supported by the UNICEF Innovation Fund, in particular with Cboard and Jellow.

The symbol sets and other resources collected on this site are intended for:

- languages that do not have localised symbols to support speech, language and literacy in health care, education and social communication situations

- all AAC users, their families and carers as well as professionals in the field interested in using symbols

- supporting low levels of literacy, learning disabilities or specific learning difficulties, where symbols can aid reading and writing skills

- helping with social interaction difficulties where symbols can act as prompts

Symbols may also work in a similar way for those with speech and language difficulties due to stroke and brain injury.

Contact us at: **hello@globalsymbols.com**

globalsymbols.com

VISION

Orbit Graphiti

- Hardware
- £££
- Employment – Education

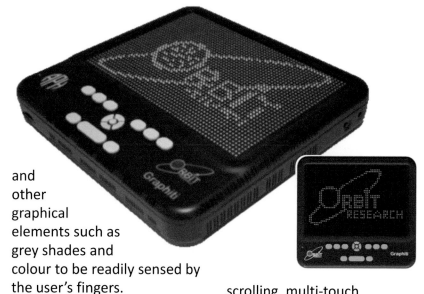

Graphiti is an Interactive Tactile Graphics Display, offering non-visual access to any form of graphical information such as charts, drawings, flowcharts, floorplans, images, and photographs, through an array of moving pins.

The technology provides the ability to set each pin to different heights, which enables the display of maps and other graphical elements such as grey shades and colour to be readily sensed by the user's fingers.

Graphiti features a touch interface to enable the user to 'draw' on the display; tracing a shape with a fingertip raises the pins along the path traced. The touch interface allows traditional forms of touch commands such as scrolling, multi-touch gestures.

Graphiti includes a Perkins-style 8-key braille keypad, a cursor pad for navigation, a standard USB host port, and an SD-card slot for loading files for reading and editing in a standalone mode.

Available at: www.orbitresearch.com

VISION

Orbit Reader and Note Taker

- Hardware
- £££
- Daily Living – Employment – Education – Social – Leisure

An Orbit Reader and Note Taker is a refreshable braille display.

It is a three-in-one device serving as a self-contained book reader, a notetaker and a braille display. It can be used as a stand-alone device or is easily connected to a computer or smartphone via USB or Bluetooth

Orbit Readers and Note Takers work with all popular screen readers across various platforms. Users can browse the Internet, read/send messages and emails, and control their computers or smartphone.

Features include:

● Book reader – read any file from an SD card. Includes features for easy navigation, inserting and editing bookmarks, browsing through folders, etc.

● Braille Display – connects to one or multiple devices across a range of platforms and uses the screen reading capability on the connected device (e.g. BrailleBack). Users can then browse the Internet, read/send messages and emails, and control their computer or smartphone.

● Notetaker – offers file creation, editing and management functions. Users can create new files and folders and save them on an SD card. They can edit, rename, delete or copy existing files

All Orbit Readers and Notetakers include calendar, clock/alarm and calculator functions. They feature onboard forward and backward translation with support for 40+ languages.

Orbit Readers and Note Takers come in two main models. The Orbit 20 and the Orbit 40, respectively, have 20 or 40 eight-dot refreshable braille cells. Both models have a Cursor pad with 4-way arrows and select keys for easy navigation. They are powered by a replaceable battery and can be recharged via USB.

Available at: orbitresearch.com

VISION DYSLEXIA

OrCam

- Hardware
- £££
- Daily Living – Employment – Education – Social – Leisure

OrCam is a company that has developed products to help people who have difficulty reading, either because of visual impairment or because they have a different need, such as dyslexia.

OrCam Read and OrCam MyEye are portable artificial vision devices that scan text and objects and then give audio feedback, describing what they cannot read or see to the user.

The OrCam Read is a handheld device containing a smart camera. Users can scan any printed text with the camera, be it on paper, book, newspaper, poster/notice or even on screen. The OrCam Read reads the text aloud as the user scans the text. Users get immediate audible feedback.

The OrCam MyEye performs similar functions to the OrCam read, but it is not held in the user's hand. Instead, it magnetically attaches to the user's glasses and is automatically directed to wherever the user is looking. As well as providing access to books newspapers etc., OrCam devices can help users go about their daily lives.

They make it much easier for users to process everyday objects such as the labels on their food shopping and medicine bottles, menus and bus timetables etc.

The OrCam MyEye can even recognise faces and relay that information back to the user.

OrCam devices are completely stand-alone – no need to be connected to the internet.

Available at: www.orcam.com

COMMUNICATION PHYSICAL COGNITIVE AUTISM

OTTAA

- Software
- App = Free Hardware = ££/£££
- Daily Living – Employment – Education – Social

OTTAA Project is an AAC (alternative and augmentative communication) system, intended for people with speech disabilities. It is a mobile, fast and effective tool that significantly improves the quality of life and facilitates social and labour integration. You can create a sentence using symbols, these are images that represent actions, objects, feelings or emotions. You can also set favourite locations and automatically get related pictograms when getting near the place.

OTTAA allows the user to communicate in different languages using preferred symbols and you can synchronise your diary better to predict the pictograms that might be needed.

You can also customise the image or text of each pictogram, also can create groups of pictograms, which enable them to create complex sentences.

You can access OTTAA with a range of devices to allow users with physical disabilities to control OTTAA.

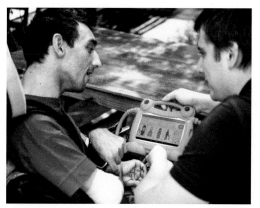

Available at: www.ottaaproject.com

VISION HEARING COMMUNICATION PHYSICAL **COGNITIVE AUTISM DYSLEXIA**

Oura Ring Generation 3

 WEBSITE

- Hardware/App (mobile)
- £££
- Daily living

The Oura Ring 3 ships to the UK from Finland. It represents a new form of wearable technology referred to as 'smart jewellery' as it takes the form of a fashionable ring you wear on a finger.

The Ring 3 has multiple sensors which connect to your app and present data about your health and well-being. Worn as a ring, it is much more likely to be usable 100% of the time as long as charged.

The ring can help you to monitor your activity, physical and mental health, and sleep. It can even monitor your heart rate and track your body temperature around the clock, as well as predict your next period.

For those who find fitness trackers difficult to wear due to physical disabilities this is an interesting option.

Available at: ouraring.com

VISION HEARING COMMUNICATION PHYSICAL **COGNITIVE AUTISM DYSLEXIA**

Pacer

- APP (mobile)
- FREE/££
- Daily Living – Social – Leisure

Pacer aims to help you to get active, lose weight, and feel better. It can track our steps 24/7 using just your phone using Pacer's interface. Beat your goals.

Pacer supports you to walk more and acts as a health coach. You can track all your activities in one app and you'll get motivation and support from the Pacer community. Pacer offers a series of fun challenges, data insights, outdoor routes, personalised fitness plans, and guided workouts to help you achieve your unique fitness goals.

No wristband or other hardware is required. Pacer works entirely from your phone with no additional setup

Pacer:
- tracks your steps all day long whether your phone is in your hand, in your pocket, in your jacket, on an armband or in your purse
- links to Apple Health to track with your apple watch
- records steps, flights, calories, distance, and active time
- uses GPS to track your outdoor walking, hiking, running, and biking on a map.

Available at: www.mypacer.com

180

VISION HEARING COMMUNICATION PHYSICAL COGNITIVE AUTISM DYSLEXIA

Passenger Assistance

- App (mobile)
- FREE
- Daily Life – Social – Leisure

Info sent straight to your train operator

Passenger Assistance is an App for National Rail which allows you to request assistance via a smartphone app. The app will help you to plan assistance for your Journey and is available on iOS and Android mobile devices.

The App allows you to:
● Request assistance for your rail journey
● Manage your customer profile
● View your travel history
● Browse rail journeys via the online journey planner
  By requesting assistance via your smartphone you do not need to contact the rail company by phone or email. You will still need to buy a ticket separately for your journey.

The app will send your assistance request for each journey to the relevant train company who will first send an acknowledgment and then follow up with a confirmation once the request has been checked.

Rail staff will then be on hand to deliver assistance, as required, throughout your journey.

Available at: play.google.com + apps.apple.com

VISION HEARING COMMUNICATION PHYSICAL COGNITIVE AUTISM DYSLEXIA

Peak

- APP (mobile)
- FREE/££
- Daily Living

Peak games are designed to push you mentally with short, intense mental workouts designed around your life.

You can challenge the skills that matter to you most with games that test your Focus, Memory, Problem Solving, Mental Agility, and others.

The games and tasks are designed to be fun, but stretch your capacity and help you to practice using skills that can deteriorate with time.

For this reason, such Brain training is especially helpful as you age.

You can compete against yourself or share scores with friends and family to encourage you further.

Available at: www.peak.net

181

VISION

PENfriend 3

- Hardware/Consumer tech
- ££
- Daily Living – Employment – Education – Social – Leisure

PenFriend 3 is an easy-to-use audio labeller designed to enable users with a range of visual impairments to live more independently.

Users speak into the in-built microphone to record their own audio labels for any object they choose. For example, it can be used to identify medicine bottles, food packets, personal hygiene items, CDs, keys, and much more. Users simply place the PenFriend onto the label, and it will play back their recording.

The dedicated labels are available in a range of sizes and colours. Special laundry labels are available for clothing. A selection of pre-recorded labels is also available.

PenFriend 3 user instructions, spoken in an English female voice, are built into labels on its packaging.

The PENfriend 3 can also be used as a Talking Book player and an MP3 music player.

Penfriend Video demonstration: www.youtube.com/watch?v=r_YdBANsoQU

Available at: www.penfriendlabeller.com

HEARING

Phonak Roger Easypen

- Hardware
- £££
- Daily Living – Employment – Education – Social – Leisure

The Phonak Roger Pen is not a piece of writing equipment but is a Bluetooth transmitter that takes its name from its pen-shaped appearance.

The pen works in conjunction with smartphones and other devices with an audio receiver. It can also transmit to a miniature receiver attached directly to the user's hearing aids or cochlear implants. This makes it a useful tool for people with a hearing impairment.

The Roger Pen is designed to capture and transmit sound over a distance of up to 20m, ideal for situations where the user is not in immediate face-to-face conversation with the person speaking, e.g. if a lecturer in a college wore the Roger Pen on a lanyard, every word would be transmitted to the user at the back of hall.

The Roger Pen can also be used in more intimate settings such as informal chats or at home to listen to music or TV.

The Roger Pen features fully automated settings and is connected with single click action. The user can listen through their smartphone earphones or other headsets. Most personal hearing aid brands can link to a miniature receiver that is compatible with the pen system.

Available at: www.phonaknhs.co.uk

COMMUNICATION

PicSeePal

- Consumer tech
- ££
- Daily Living – Employment – Education – Social

PicSeePal is an AAC (alternative and augmentative communication) housing or case that is lightweight, portable, customisable, splash-proof, modular, and easy to use.

It offers a low-tech solution that you can be taken anywhere to enable AAC users.

The case houses a number of printed communication boards that can then be used to communicate when an electronic device is not available.

It can used by all ages. For some people it is an ideal system for communication in itself, for others, it provides a backup system for challenging circumstances.

Available at: picseepal.com

VISION

PLEXTALK Pocket

- Hardware/Consumer tech
- £££
- Daily Living – Employment – Education – Social – Leisure

The PLEXTALK Pocket is a portable-sized book reader and recorder. Users can use it to playback existing digital talking books.

It also has a wireless LAN capability that can be used to download or stream Web Radio and Podcasts.

The PLEXTALK Pocket is a useful tool for anyone with a visual impairment, especially in an educational or work setting.

Users can employ the record function to record memos and short notes to themselves and make longer recordings. For example, it could record a meeting or students can use the PLEXTALK Pocket to record lectures for later review. A useful feature when recording is the ability to insert headers using a simple key press.

Other Features Include:
- Large tactile keys
- Voice guides for all key operations and menu items
- Sleep timer
- 10 Hour playback
- 8 hour MP3 recording
- Daisy editable software

- CD playable with an external USB drive

NB: As of July 2020, PLEXTALK Pocket users are no longer able to access content from Audible.

Available at: www.plextalk.com

> HEARING

Portable Induction Loop

- Hardware
- £££
- Daily Living – Employment – Education – Social – Leisure

An Induction Loop System amplifies sound for the benefit of hearing aid users. The Loop uses a magnetic field (created from a wire loop) to transmit a signal. Anyone in the area using hearing aids can pick up the transmission. The sound is amplified, but it is also clearer as any background noise is reduced or cut out altogether. The hearing aids must be switched to the T position to access an Induction Loop transmission.

It is common for Induction loop systems to be permanently installed in venues such as theatres, cinemas, places of worship, public service areas, etc. However, sometimes this is not the case. If a user is visiting a site with no existing Induction Loop system, it can be beneficial to have a portable, personal induction loop.

A Portable Induction Loop is intended for face-to-face conversations. It consists of a small portable, battery-powered amplifier placed directly between the user and the person speaking. The in-built microphone picks up what the speaker is saying and transmits it to the user's hearing aids. Because they are portable, they can be used in many situations, such as medical settings, at home, school or the workplace.

VISION HEARING COMMUNICATION PHYSICAL **COGNITIVE** **AUTISM** **DYSLEXIA**

Prizmo

- Software/App (mobile)
- ££
- Daily Living – Employment – Education – Social – Leisure

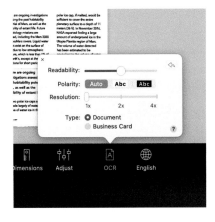

Prizmo for iOS and Mac is a scanning app with built-in optical character recognition. (OCR) It works with pictures taken on the user's smart device or digital camera, screen shots, and documents transmitted from connected or wifi-enabled scanners. No matter the source, Prizmo's efficient editing tools and built-in OCR will help the user access and process the document.

Prizmo is easy to use and includes features to assist the user in taking a usable photo. It can correct perspective and uses a grid to straighten the image. Its Flatten Curve option means that book and magazine pages are easily captured.

After processing OCR on a document, a user can have the text read out in various voices and speeds. (The app has a choice of over 90 male/female voices covering a range of 26 languages). Prizmo features synchronised highlighting so a user can follow along as the text is readout.

This feature is of particular benefit to users with dyslexia or a similar barrier to reading. There are navigation settings enabling play and pause or skipping to the next section.

Available at: creaceed.com

COMMUNICATION

Proloquo2Go

- App (mobile)
- £££
- Daily Living – Employment – Education – Social – Leisure

Proloquo2Go is an AAC (alternative and augmentative communication) app for iPhones and iPad.

It is designed to be used as a daily communication tool for users who face a range of barriers to verbalisation. Obstacles such as autism, cerebral palsy, aphasia, apraxia or other speech impediments.

Proloquo2Go can help users

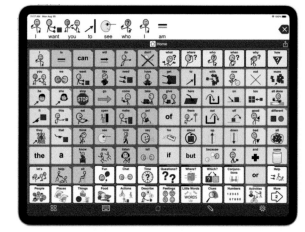

build language skills and express their thoughts, emotions, wants, and needs. It has been developed to be accessible to users with a range of visual and fine motor skills and is completely customisable to the user.

Proloquo2Go features core words to assist users in building sentences. As users gradually build their vocabulary, there is no limit to how far they can progress.

In addition to the core words, there is a fringe word vocabulary of over 10,000 words and customisable vocabulary levels.

Proloquo2Go has over 100 natural-sounding text-to-speech voices in a range of languages. Its accessibility settings fully support VoiceOver and scanning features.

Available at: www.assistiveware.com

COMMUNICATION

ProxTalker

- Hardware
- £££
- Daily Living – Employment – Education – Social

The Logan ProxTalker is a recorded speech communication device available. It uses RFID (radio frequency identification) technology to enable independent verbal picture communication for non-verbal people of all ages. You place any photo, symbol or object onto a sound tag card and give it a voice. To trigger speech you place your sound tag card on any one of the five buttons and push.

A standard package includes: 80 prerecorded sound tags, 100 blank small tags, carry case with 4 page velcro set, 2 sets of programming tags, 4 colour pages with velcro strips, peel-and-stick labels and laminates, tool kit.

The ProxTalker is:
- robust, water-resistant, and easy to use

- portable in backpack or wheelchair/wall mountable in binder format
- offers a light touch option for easier access
- tags can be labeled with a wide range of symbols.

Available at: logantech.com

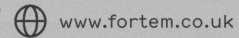 www.fortem.co.uk

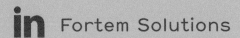 Fortem Solutions

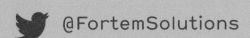 @FortemSolutions

WE ARE FORTEM

Established in 2002, Fortem provides specialist property solutions tailored to keep homes and buildings running smoothly. Our highly trained teams deliver a range of internal and external repairs 24/7 as well as high quality voids, planned installations and gas services.

Our experience, commitment to best practice and exemplary quality, together with a genuine partnered approach, ensures we achieve excellent outcomes for our partners, clients and customers. We have an established track record of improving communities and creating better places for people to work and live.

SCAN ME!

FOR THE LATEST FROM FORTEM

COGNITIVE AUTISM DYSLEXIA

Pushover

 WEBSITE

- Software/App (mobile)
- FREE/£
- Daily Living – Employment – Education – Social

Pushover allows you to receive unlimited notifications on all of your devices from dozens of websites, services, and applications that are integrated with Pushover.

You have to provide your Pushover User Key or e-mail address and all your notifications will appear in one place.

Individuals can use Pushover for Android, iOS, and Desktop for a one-time in-app purchase on each platform that they need, following a 30-day free trial. After the one-off payment is made there is no subscription fee.

Pushover integrates with web apps like IFTTT, security cameras, IoT devices, and many other things that send alerts to your phone, tablet or computer.

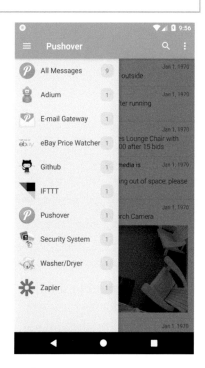

Available at: pushover.net

VISION

Ray Vision App

- App (mobile)
- FREE for 1st 30 days / then ££
- Daily living

The Ray Vision App by Project Ray is an Android designed to make smartphones and tablets accessible to users with visual impairments.

Its vision-free interface employs voice recognition, simple gestures, tactile touching, and audio feedback to replace the standard vision-dependent smartphone controls. The app also allows users to change contrast and colour settings on their phones to suit their own individual needs and preferences.

As well as increasing accessibility to the user's phone, the Ray Vision App also offers 'Over the Web' assistance. This provides remote aid for device discovery, navigation, and visual assistance.

All the app's features can be accessed with one hand and include voice-operated speed dial, call history, text-to-speech, voice SMS, messaging services, visual recognition, OCR, calendar and clock.

Project Ray NFC Stickers affix to the back of a smart

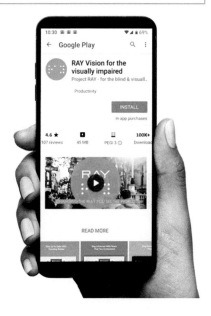

device to be used as tactile buttons to control pre-set actions in the Ray Vision app.

Project Ray also sells a range of mobile phones designed for users with a visual impairment, but these are currently only available in the USA.

Available at: project-ray.com

VISION

QCast

- Software
- Demo version = Free Licence = £
- Daily Living – Leisure – Education

QCast is designed to simplify access to podcasts for users with a visual impairment via their Windows PC or Mac computer.

Most users access podcasts through a smartphone or tablet, but QCast is an intuitive and accessible interface for users who need to listen to their podcasts via PC or Mac. As a rule, iOS and Android devices use podcatchers to automatically download new episodes of selected podcasts.

However, this software generally does not work well with screen readers on PCs and Macs, so it is difficult to navigate if a user has a visual impairment. With QCast, users can access all the usual features of podcatcher software, such as the automatic download of new episodes, the ability to fast forward, rewind or adjust playback speed; automatically resume listening where they previously left off.

Once a user has installed QCast, they have the option to import their podcasts from other software and can access podcasts from all over the world, even password-protected ones.

Available at: getaccessibleapps.com

PHYSICAL

Quadstick

- Hardware
- FREE
- Leisure

Quadstick describes itself as 'a game controller for quadriplegics'. It is a hands-free control to enable players with profound physical disabilities to play games on consoles and PCs .

The Quadstick is available in three different models ad a variety of mounting arms.

The Quadstick Singleton has a joystick and a sip/puff sensor. It is designed to be used to operate a personal computer rather than a games console.

The Quadstick Original has a joystick, 4 sip/puff pressure sensors, and a lip position sensor. The latter can be assigned to any mouse movement button, keyboard key or game controller button.

The Quadstick FPS is the same as the original but with a higher quality joystick gimbal. Also, with this model, players can customise mapping between the inputs and outputs and quickly change them even while playing a game.

All models of Quadstick can connect directly to a PC/Mac, Nintendo Switch, PS3 or PS4 but will require an adapter to connect to Xbox One or 360.

Available at: www.quadstick.com

PHYSICAL

Quha Zono Mouse

- Hardware
- £££
- Daily Living – Employment – Education – Social – Leisure

Quha Zono is a lightweight gyroscopic mouse that enables accurate and intuitive computer access with small body movements.

It is designed for users who cannot operate a computer mouse with their hands. It can be attached to any part of a user's body according to their individual needs. As long as the Quha Zono is in front of the device being accessed, any small movements by the user will be sufficient

Quha Zono is completely wireless and compatible across various devices and platforms. The included adapter allows for the connection of two switches. Or one switch can be used via a 3.5mm stereo connector. When fully charged, the Quha Zono has a use time of up to 30 hours.

Quha Dwell Software is available to purchase separately and can be used to tailor the mouse more accurately to the user's needs.

The enhanced Quha Zono2 mouse is also available.

Available at: www.quha.com

COMMUNICATION

QuickTalker Feathertouch 7

- Hardware
- £££
- Daily Living – Education – Social

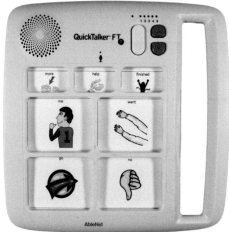

The QuickTalker FT7 is a portable, battery-operated, multi-message communication device. It is a useful tool for users who face a range of communication challenges, both cognitive (such as autism) and physical.

Each of its seven locations can be linked to a recorded message. Messages can be full sentences or single words that make up a sentence when pressed in the correct order: "I…. want…. a…. drink."

Symbols to represent each location are printed on an overlay and inserted under the feather touch membrane that covers the device – so-called because it responds to a very light touch from the user.

Overlays are not included, but personalised overlays to fit the QuickTalker can be designed using the Ablenet Symbol Overlay Maker (available at apps.apple.com).

The unit offers up to six minutes of recording time.

Retailers include:
www.inclusive.co.uk
www.liberator.co.uk

Also available are QuickTalker FeatherTouch 12 and QuickTalker FeatherTouch 23.

Info at: www.ablenetinc.com

VISION

Refreshable braille displays

- Hardware
- £££
- Daily Living – Employment – Education – Social – Leisure

Refreshable braille displays are tactile devices that enable users to navigate and read information in dynamic braille.

Permanent braille comprises braille dots (a series of holes or bumps on the reading surface). With a refreshable braille display, these dots are represented by small pins that are electronically raised or lowered through six holes, each representing a braille cell.

The display is linked to a computer, smartphone or similar and text on the screen is converted into braille on the refreshable braille display. The number of refreshable braille cells on a device varies according to make and model. Users read by moving their fingers across the cells as if they were reading braille on paper.

Refreshable braille displays come in a range of sizes. They can be as large as a standard keyboard for a desktop computer. Or they can be smaller, more portable models designed to be used with smartphones and tablets.

Unlike hard copies of braille documents, refreshable braille displays have the advantage of storing a large amount of information on a small device.

Users with a visual impairment can use the refreshable display to read text output. They are not limited to existing text. For example, if a user employs voice recognition software to create their own documents. They can use a refreshable braille display to quietly read it back to themselves and proofread what they have 'written'.

Some refreshable braille displays also have note-taking functionality, while some incorporate a 'Perkins-style' keyboard which can be used to control the computer.

Sample products:
shop.rnib.org.uk/mantis-q40-braille-display-90717
blitab.com/
store.humanware.com/heu/blindness-braillenote-touch-plus-32.html

HEARING   COMMUNICATION   AUTISM

Relay UK

 WEBSITE

- Software/App (mobile)
- FREE + normal call charges
- Daily Living – Employment – Education – Social – Leisure

Relay UK is an app designed to assist users with hearing and/or speech impairments to make calls on their smartphones.

A Relay Assistant (RA) acts as a go-between for the user and the person they are calling. The user can speak into the phone or type what they wish to say.

The RA will listen to the other person then type their response so the user can read it.

This system works in real-time for immediate responses.

When users install the Relay UK app on their phone, they will link their phone number to the app.

Whenever they open the app before dialling a number, the app will automatically include a prefix of 18001, which will direct the call via the Relay UK service. The user presses 'Join' on the app.

Then when the call is answered, an RA will introduce the call to the recipient and, where necessary, explain how the service works.

The RA will type in the other person's side of the conversation. When the call is finished, the user hangs up as normal and then disconnects/closes the app.

Relay UK can also be used for 999 emergency calls. This works in the same way as any other call. If the user prefers to type their side of the conversation, there is the option to select from a list of favourite phrases to increase typing speed.

Although Relay UK is primarily available as a smart device or website app, users can also access it on text phones.

A user Simply Dials 18001 followed by the phone number they wish to call. (Including the dialling code). The textphone screen will display messages to inform the user if the call has been answered and the status of the Relay UK service. Once connected to a Relay UK assistant, the user can type their side of the conversation.

The Relay UK app is a free service. The user simply pays their normal calling charges.

Windows PC users can Get Relay UK from the Microsoft Store.

Available at: www.relayuk.bt.com

COGNITIVE AUTISM DYSLEXIA

Remember The Milk

- App (mobile)
- FREE/££
- Daily Living – Employment – Education – Social

Remember The Milk is a to-do app to help you organise your daily life. It helps remind you of tasks you need to do:
- offers reminders via email, text, Twitter, and mobile notifications
- you can share your lists and give tasks to others to get things done faster
- stay in sync on all of your devices
- organise the way you want to with priorities, due dates, repeats, lists, tags and more
- search your tasks and notes, and save your favourite searches as smart lists
- integrates with Gmail, Google Calendar, Twitter, Evernote and other apps.

Remember The Milk is free to download and use. Purchasing a Pro subscription in the app unlock features such as:
- subtasks – break your tasks down into smaller, more manageable pieces
- unlimited sharing – share your lists with others to get things done faster
- attach files to your tasks – connect to Dropbox or Google Drive
- unlimited storage – keep track of all your hard work with unlimited completed tasks.

Available at: www.rememberthemilk.com

MENTAL HEALTH

Revere

 WEBSITE

- App (mobile)
- FREE or subscription
- Daily Living – Social

Revere is an app designed to help people remember people.

Users can store almost any information about the people in their lives, ranging from their birthday to their work details to whether they take sugar in their tea. The app is designed to help users remember people-focused information and use it to build better relationships with those around them.

A user can use voice capture to quickly record notes about the people they have listed in the app. It is a useful tool to help remember details about people they will meet again, such as how they met, what their children are called or even what the last conversation was about.

The app can also set reminders for recurring events such as birthdays or single events such as medical appointments and job interviews.

People's names can be imported from contacts and a file built up for each person. This can consist of both dedicated fields for names and background and the option to include freeform notes as needed.

NOTE: users must create an account with an email address to back up and sync notes to all their devices.

Chrome extension to access Revere online is only available with the paid subscription.

Available at: www.revereapp.com

VISION COGNITIVE AUTISM DYSLEXIA

RoboBraille

WEBSITE

- Software
- FREE/£/££/£££
- Daily Living – Employment – Education – Social – Leisure

RoboBraille is an email and web-based service capable of automatically transforming documents into a variety of alternate formats for the visually and reading impaired.

RoboBraille is accessible 24/7 as a self-service solution and is available free of charge to an individual, non-

commercial user, who is not affiliated with an institutional setting that is obliged to provide support (academic institution, organisation or association). You do not need to register to use the service. The objective is to support and promote the self-sufficiency of people with special needs socially, throughout the educational system, and in the labour market.

You upload your original file in one of a range of formats including Word, PowerPoint, PDF, ePub or even some image files, wi9thin a few minutes a converted file is sent to you. You can choose from a braille-ready file, mp3, Daisy, ePub, and Mobi formats for the file.

Available at: www.robobraille.org

HEARING

Rogervoice

- App (mobile)
- FREE/£/££
- Daily Living – Social

Rogervoice is an app that converts phone calls into captions on the user's smartphone screen.

It works in real-time, so the user can instantly see what a caller is saying. This makes it ideal for users with hearing impairments, including those who simply struggle to hear when making/receiving calls in a noisy environment.

No special equipment is needed, only a smartphone with 4G or wifi. When a Rogervoice app user makes a call, the recipient's voice is analyzed by the app's algorithm and transformed into text.

Users can also receive calls – they are given a special Rogervoice number to share with other people. When phoned on that number, the app picks up the call.

The free app only lets users call other people using the

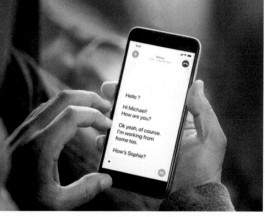

Rogervoice app. If a user wishes to use the app when calling a mobile or landline, they need to upgrade to one of the monthly subscriptions.

The app-text/ background is colour-customisable according to the user's needs. Only one side of the conversation is transcribed, but transcripts of a call can be saved to re-read later.

Available at: rogervoice.com

PHYSICAL
Rollerballs

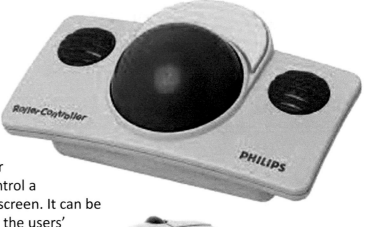

- Hardware
- ££/£££
- Daily Living – Employment – Education – Social – Leisure

Rollerballs or trackballs are alternatives to using a keyboard, mouse or joystick.

They allow the user to carry out very precise movements, so they are a favourite of graphic artists and animators.

But they are also a useful tool for people who face challenges with fine motor control as they provide a greater control surface.

Rollerballs usually consist of a housing containing a ball in a socket.

This can be moved or rolled to control a cursor on a screen. It can be operated by the users' hand, arm, elbow or foot, suitable for people with a range of physical disabilities.

Rollerballs and trackballs come in a range of makes and sizes to buy as an addition to an existing computer or device.

MENTAL HEALTH

SAM (Self-help App for the Mind)

- App (mobile)
- FREE
- Daily Living

SAM offers a range of tools to help users monitor their moods and feelings; techniques to help them recognise triggers and challenge and change negative thoughts and emotions; and a social cloud feature that enables them to receive and support other users.

The app's guided self-help techniques are organised into themes such as anxiety or loneliness. These include activities for immediate effects, such as breathing techniques and other techniques and games to build resilience going forward.

The mood tracking tools encourage the user to record their feelings on a scale and provide a visual summary to help them monitor changes over time.

Use of the Social Cloud feature is entirely optional. The developers ask users of this feature to be non-judgemental and sensitive in their interactions with other users.

NOTE: SAM does not offer clinical diagnoses or therapy programmes, although it provides relevant links for these and contacts for more immediate help.

Available at: mindgarden-tech.co.uk

VISION

Seeing AI

- App (mobile)
- FREE
- Daily living – Employment – Education – Social – Leisure

Seeing AI is an app designed for users with a range of visual impairments to assist them in visualising the world around them.

The app uses a device's camera and artificial intelligence or AI to describe many aspects of the user's immediate environment.

Seeing AI can be used to read text aloud, recognising both printed text and handwriting. It can identify different banknotes and announce what it 'sees'.

The app also includes a barcode scanner which can be used to identify products.

However, as well as objects, the Seeing AI app can use facial recognition to scan nearby people and pick out the user's friends.

It will describe unknown people by their physical appearances, such as age and gender, and by their expressions and emotions.

It also has a currently experimental, Scene feature, which tries to describe the scene around the user.

Seeing AI can describe colours and generate an audible tone to correspond to the levels of brightness in the user's surroundings.

Info at: www.microsoft.com/en-us/ai/seeing-ai

VISION

Seeing Assistant Magnifier

- App (mobile)
- FREE
- Daily Living

The Seeing Assistant Magnifier is designed to help people with a range of visual impairments easily read small text or look at small items.

It allows the user to employ their phone as an electronic magnifying glass.

There are settings to adjust the displayed image's contrast, brightness, and colour, including the option to invert black and white to make the text easier to read.

The Seeing Assistant Magnifier allows up to 10x magnification. The app is good for magnifying and reading small print, etc.

The app can also use a front-facing camera. The Seeing Assistant Magnifier causes the user's phone to function as a magnifying mirror. This can be a helpful tool to, for example, assist users in applying make-up.

Available at: seeingassistant.tt.com.pl

Cost of living support available now

Lambeth Council and local charities are here to help during the cost-of-living crisis.

For further information visit our website
lambeth.gov.uk/cost-living-crisis-support

SCAN ME

We can help you with:

- Council tax support and rebates
- Access to benefits and grants
- Make your home energy efficient, reduce your bills
- Access to Foodbanks and low-cost food
- Household Support Scheme - emergency help with food and fuel bills
- Job and business start-up support
- Holiday activities

Household Support

The household support scheme helps Lambeth Residents that are facing hardship, a crisis, emergency, or disaster. This includes struggling to pay bills or afford food.

We can help with vouchers for food or high street shops, fuel payments, handyman services, removals and storage along with physical goods such as refurbished white goods and second-hand furniture.

Post Office payout

Lambeth households who do not pay their council tax via Direct Debit should have received a letter via the post to claim their council tax energy rebate, simply take this letter with a form of ID to any Post Office branch. Households in receipt of Council Tax Support will have received an additional payout.

COMMUNICATION PHYSICAL COGNITIVE AUTISM DYSLEXIA

Sentence Key
WHOisDOingWHAt

- App (mobile)
- £
- Daily Living – Employment – Education – Social – Leisure

Sentence Key WHOisDOingWHAT is a sentence-creating game that is suitable for people with various educational and learning needs, including autism, language disorders and AAC (alternative and augmentative communication) users.

The user is provided with a picture such as people eating ice cream and then selects four symbols or images in the right order to form a sentence to describe what is happening.

If they create the sentence correctly, they are rewarded with a short animation. The app includes 72 possible animated sentences, but there is also the option for users to create their own sentences.

Sentence Key is designed to work on iPhones and iPads. The app is user-friendly with no in-app purchases, ads or links, and there are other apps in the same series: Sentence Key Chores, containing 91 new sentences and animations, Sentence Match, which helps to recognise sentence structures, plus the linked Sentence Match Chores.

Available at: computerade.com

MENTAL HEALTH

Serenita

- App (mobile)
- FREE or subscribe to Pro Version = £ per month or ££ per year
- Daily living

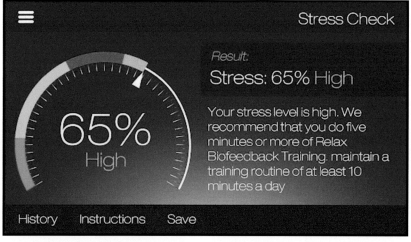

The Serenita app claims that its users can lower their stress levels by using it for just five minutes a day.

Reduced stress levels can improve other areas such as anxiety, sleep and depression. By monitoring a user's stress patterns over time, Serenita aims to give them the means to increase stress resilience.

As well as tools to monitor their stress level daily, a user can access targeted exercises designed to give them the largest reduction in stress per time spent.

Serenita uses digital technology to analyse a user's blood flow and variances in their heart and breathing rates, then uses this information to calculate their stress level.

The user simply needs to place their index finger over the back camera lens on their phone. The app will then instruct the user in a breathing routine personalised to them.

Available at: eco-fusion.com/serenita

HEARING

Signolux Alerting System

- Hardware
- ££
- Daily living

The Signolux Alerting System is suitable for users with hearing impairments. It consists of two transmitters: an audio transmitter that uses a microphone to detect sounds and a direct transmitter that broadcasts to any connected device.

When the audio transmitter detects a sound such as a doorbell, the direct transmitter will broadcast an alert. Alerts can be audible, visual or physical (vibrations) depending on the user's needs/devices available.

Receivers available include wall-mounted units, a plug-in strobe light receiver, and portable pagers. All receivers have 6-8 coloured icons to enable users to visually identify the sound source. Alerts can be loud sounds, a flashing white light or vibrations.

The wire-free system includes alerts for bell push, smoke alarms, and person-to-person. Up to 8 transmitters can be linked to each icon with 15 sounds available, each reaching up to 90dB.

Available at: www.humantechnik.com

HEARING

SignVideo

  WEBSITE

- Software/Hardware/App (mobile)/Content/Consumer tech
- £££
- Daily Living – Employment

SignVideo is a British Sign Language (BSL) Interpreting company that offers BSL/English translation services, remote Interpreting via video link (VRI), and Video Relay Services (VRS). They can provide fully qualified BSL interpreters 24 hours a day on demand.

The SignVideo services Provide support for users both in the workplace and at home. Their SignDirectory lists 150+ businesses and services such as banks, health centres, helplines, etc. Users can contact any listed companies through a SignVideo interpreter for free, whether work-related or personal.

Users who subscribe to the SignVideo service are provided with their own account. They can use this to make VRS calls, participate in meetings via VRI, and use the translation service. They can make free video calls to another SignVideo subscriber and connect calls with just one click.

SignMail allows users to watch video messages from their contacts when away from home or the office, and their personal History allows them to keep track of incoming, outgoing, and missed calls

The SignVideo work package can be funded through the Access To Work Scheme.

Available at: signvideo.co.uk

PHYSICAL

Sip-and-puff switches

- Hardware
- £££
- Daily Living – Employment – Education – Social – Leisure

Sip-and-puff or sip 'n' puff switches (SNP) are so-called because they translate the users' sips and puffs of air into independent switch closures.

They are also known as pneumatic or breath-activated switches. They are primarily intended as a tool for users who have limited or no motor capability to operate switch activated devices – users with conditions such as muscular dystrophy, multiple sclerosis or cerebral palsy, and those who are experiencing challenges resulting from spinal cord injuries, brain injuries or amputation.

SNPs enable users to access and control computers and other important technologies such as augmentative communication devices, environmental control systems or wheelchair navigation systems.

A basic SNP consists of a headframe that supports a metal or plastic tube that positions within easy reach of the user's mouth. Alternatively, the tube could be supported by a stand that clips to a solid base, such as a desk. The user operates the

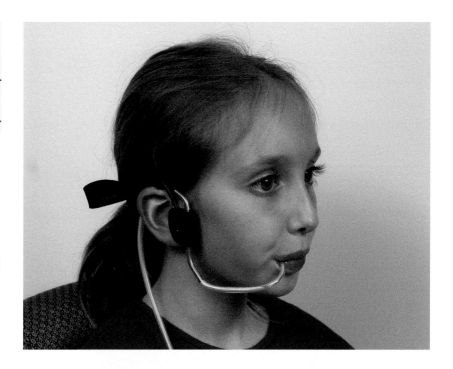

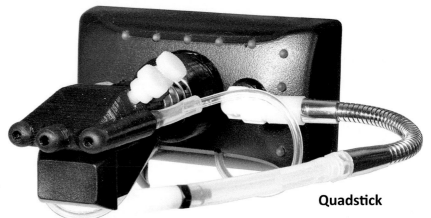

Quadstick

SNP by using the tube to take small inhalations of air (sips) or to blow/exhale air (puffs). Each sip, puff or combination can activate a switch action that allows users to access functions in a connected device. There might be a wired connection to the end device or it is possible to use a Bluetooth Switch Interface to connect an SNP to a phone or other smart device. This opens up access to various apps and other functions such as calls and messaging services.

Video: 'Sip-n-Puff Technologies for Everyday Function' at www.youtube.com/watch?v=kKYGI2NawKQ

Available at: www.liberator.co.uk • www.orin.com • enablingdevices.com

VISION HEARING COMMUNICATION PHYSICAL **COGNITIVE AUTISM DYSLEXIA**

Siri

- Integrated
- FREE
- Daily Living – Employment – Education – Social – Leisure

Siri is a digital personal assistant found on all Apple devices. It works across every Apple product, so a user's devices are linked and it can assist the user in various ways.

For example, users can ask Siri to answer questions about the news, weather or virtually

anything. Siri will search online and find an answer. It can carry out a command on a user's device, such as setting an alarm, opening an app or file or finding a photo. If a user has an environmental control app on their device, Siri can be used to issue commands such as "open blinds" or "switch off hallway light".

The Accessibility Assistant Shortcut can be used to create a customised list of recommended accessibility features based on individual needs. Users then can use the shortcuts to turn the accessibility features on or off as required.

Siri is usually voice-controlled, however, users with speech or hearing impairments can use their onscreen keyboard to interact with Siri.

Available at: apps.apple.com

PHYSICAL

Skoog Music

- Hardware
- £££
- Leisure

Skoog and Skwitch are easy to play musical instruments used in conjunction with iOS devices.

The Skwitch clips directly onto the user's iPhone, whereas the Skoog connects to an iPad via Bluetooth. Both instruments feature large buttons that the user presses to play music.

This button design makes Skoog Music instruments accessible to users who face physical or cognitive barriers to playing more traditional instruments. The buttons can be pressed with any part of the user's body, such as hands, feet or elbows, so fine motor skills are not required.

The Skoog, in particular, is an accessible instrument. It is a large cube with pressable squeezable buttons on five of its faces. Each button produces a different sound or

effect.

The free Skoog app allows users to choose songs from their Spotify or Apple Music playlists, then auto-tunes the Skoog so they can play along. The Skoog can also be used with other apps, such as GarageBand.

Available at: skoogmusic.com

COMMUNICATION PHYSICAL **COGNITIVE** AUTISM

Skyle for Windows

- Software/Hardware
- £££
- Daily Living – Employment – Education – Social – Leisure

Skyle for Windows consists of the Skyle for Windows Eye Tracker used with the Skyle EyeMouse software.

Together they enable users at school and at home to use their gaze to communicate, play games, and join in classroom activities to help them explore and learn.

The Style Eye Tracker is easy to set up and simple to use. It is also fully portable and can easily be moved between computers/locations.

It can work in most lighting conditions but does require the associated screen to be sized between 10-24 inches.

The Eye Tracker features a 'No-Fail-Start' one-point calibration setting, which means that users of any ability level can get started immediately and gradually develop their eye-gaze skills at a rate that suits them. For more practised users, up to a nine-point calibration is available.

The Skyle Eye Tracker can track where the user's eyes are looking and uses that information to control the movement of the onscreen mouse cursor.

The Skyle EyeMouse then allows the user to perform any actions carried out by a standard mouse – Left/Right Click, Double Clicks, dwell, etc.

Individual profile settings and customisation options mean that the program can be calibrated to cater to the specific needs of individual users.

Skyle for Windows is suitable for use with both children and adults. It is a useful tool for users experiencing a range of physical, intellectual, and communication difficulties. For example, users with cerebral palsy, multiple sclerosis, a spinal injury or autism.

Demonstration video: www.youtube.com/watch?v=aJGOQW29zqo

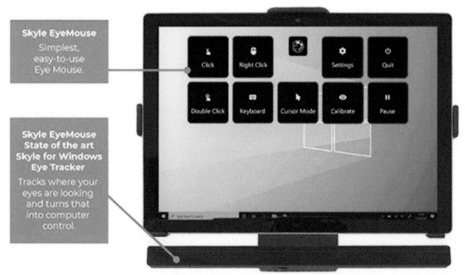

- Easy set-up: **No technical training or knowledge required.**
- No-Fail-Start: **1 point calibration for early users.**
- Unique Classroom Features: **Designed for special needs students and teachers.**
- Building Independence & AAC: **With Skyle for Windows learners can develop at their own pace.**

Available at: www.inclusive.co.uk

COMMUNICATION PHYSICAL AUTISM

Smartbox – Eye Tracking

- Hardware
- £££
- Daily Living – Employment – Education

Smartbox offers a range of cameras and peripherals that enhance their Grid 10 and Grid 12 communication devices by adding eye tracking capacity.

These include the Lumin-i eye tracker for AAC (alternative and augmentative communication), the Alea IntelliGaze eye tracker an effective eye gaze camera, and the Eyegaze Encore a camera that can predict your gaze point to 1/4 inch or less.

The camera is designed to track just one eye when needed with minimal infrared light, making it a good option for people with ventilators which can confuse other cameras that shine more infrared light across the whole face.

Smartbox also offers Eye gaze software which is fully eye gaze compatible. This includes Look to Learn exploring visual scenes and playing games.

Grid 3 gives you control of a system, with eye gaze integration with natural features such as zoom to click.

The solutions are effective and reliable indoors and outdoors.

Available at: thinksmartbox.com

MENTAL HEALTH

Smiling Mind

  WEBSITE

- App (mobile)/Content
- FREE
- Daily Living

Smiling Mind is an app that aims to introduce its users to mindfulness and thus 'bring balance' into their lives.

The app provides guided meditations. It suggests ten minutes use a day to reduce anxiety, increase relaxation and concentration, promote better health and sleeping patterns and much more.

Smiling Mind has programmes for all ages from 3 years upwards and is designed to be followed anytime.

Available at: www.smilingmind.com.au

VISION **HEARING**

SoundALERT

- App (mobile)
- FREE
- Daily living

The SoundALERT app turns any smart device into a high-tech alerting device.

SoundALERT uses the device microphone to pick up sounds from the user's environment and then converts them into a visual/vibrating alert.

The alerts take the form of onscreen notifications, vibrations, and flashing lights.

The Sound ALERT app can pick up a wide range of sounds. Some smoke, carbon monoxide and gas alarms are pre-installed on the app. Others can be added by the user according to their own appliances, e.g. doorbells, intercoms etc.

The SoundALERT app does not need to be connected to the internet. It enables users with a range of hearing impairments to feel safe and secure.

NOTE: Fully compatible with common brands of alert products such as Bellman and Geemarc.

Info at: www.soundalert.co

HEARING

Sound Amplifier

- App (mobile)
- FREE
- Daily living – Social

Sound Amplifier is an Android Accessibility app that helps users with hearing impairment hear more clearly. It works when a user connects their headphones and, using the app, can customise the frequencies of the sounds they can hear.

This means they can enhance important sounds, like voices, while at the same time filtering out distracting background noise. Because users adjust the settings themselves, Sound Amplifier works for every user according to their individual needs.

While adjusting the settings, some users may find it difficult to ensure that Sound Amplifier is detecting or augmenting sound. To address that issue, Sound Amplifier has an onscreen audio visualisation feature. It works a little like a volume control to show users how much they are boosting a sound.

Once installed, users can easily launch the app directly from their home screen without opening the accessibility settings every time they want to use it.

NOTE: Sound Amplifier can be used with both wired and Bluetooth headphones.

Available at: play.google.com

203

PHYSICAL

Spec Switch

- Hardware
- ££
- Daily Living – Employment – Education – Social – Leisure

The Spec Switch is a wearable or mountable assistive switch. It can be used by people with motor impairments to access and control technology, such as communication devices, computer technology and environmental controls.

The Spec Switch is a wired, smooth plastic switch. It is fitted with a 1.8m cable and is small enough to be worn with only a 3.5cm diameter activation surface. However, it is most often used as a mounted switch and is supplied with a 60cm strap for securing it around items such as headrests.

It comes with a flush base allowing the switch to be firmly screwed to a flat surface, such as a desk.

The Spec switch is available in five different colours, making it easy to distinguish between different switches assigned to different purposes.

Available at: Adapt-IT.co.uk

VISION HEARING COMMUNICATION

Streamer

 WEBSITE

- Software/App (mobile)
- 30 day free Trial
  Annual subscription = ££ per year
- Daily Living – Employment – Education – Social – Leisure

With Streamer, users can create a personal, secure website that captions and translates communications.

It can share documents, add notes to transcripts, send private messages, and even simultaneously translate conversations into multiple languages. Everything is confidential, with data and transcripts belonging to the user alone.

For users who experience barriers to hearing, Streamer is a way to access both in-person communications (face-to-face conversations, lectures) and onscreen events (videos, live streams).

The software captions and/or translates (users have a choice of open or closed captioning or both). It is then easy to annotate the live transcripts with colour-coded notes.

For Streamer users who are non-verbal, the 'PhraseBuilder' tools will speak their sentences in a personalised voice. There is the option to use pre-set phrases or users can create their own core sentences.

Available at: streamer.center • apps.apple.com

HEARING

Subtitle Viewer

- App (mobile)
- FREE
- Leisure

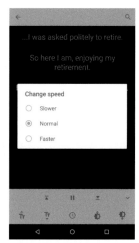

Subtitle Viewer is an Android app designed to find, download and display subtitles on a user's smartphone.

Users can search for subtitles linked to a specific film or episode of a show. Once downloaded, the subtitles are synchronised to what the user is watching and played in real-time. The subtitles are slowly scrolled up the screen, highlighting the current phrase or sentence.

Because Subtitle Viewer works on a smartphone, it is fully portable. Users can view the subtitles wherever they can view the film; at home, in the cinema, on a train, etc.

Subtitle Viewer is a useful tool for any user with a hearing impairment.

Please note: Subtitle Viewer does not play films or shows. It only displays the subtitles.

Info at: www.subtitleviewer.com

VISION

SuperNova Magnifier and Screen Reader

- Software
- FREE/£££
- Daily Living – Employment – Education – Social – Leisure

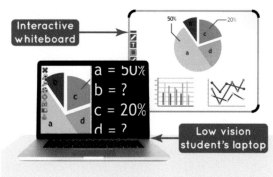

Supernova is a combination magnifier, screen reader, and braille reader for Windows.

It is a particularly useful tool for users with a range of visual impairments. As a magnifier, it enlarges and enhances parts or the whole of the user's screen.

It will read aloud characters and words as the user types. Or it can be used to read and describe documents, web pages, emails, etc. There is also a scan and read option for paper documents or otherwise difficult to access PDFs. Supernova supports a wide range of braille displays.

The program's features are accessed via the SuperNova Control Panel, and include human sounding speech and braille Display Support – over 60 models supported.

Info at: yourdolphin.com

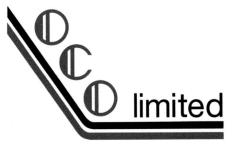

OCO Limited is a family run business established in 1979, we specialise in the maintenance, repair, servicing, and installation of heating, mechanical, electrical and water hygiene building services and equipment in the private and commercial sectors.

We work in domestic, communal and commercial environments, covering everything from roof level to below ground projects.

We also have an in-house Aids and Adaptations department managed by an industry specialist.

We pride ourselves on providing a reliable, professional and high quality service to all our loyal and valued clients and customers.

At OCO Limited we're proud to help and support the clients & communities we serve. We've made a commitment to safeguard their futures through employment and sustainability initiatives.

Social Value is at the heart of all our operations, and we offer the following commitments to all the communities we provide services for:

- Local apprenticeship scheme
- Local supply chain
- Work experience programme
- Local company mentor programme
- Estate sponsorship, including regular social visits and aid for sheltered accommodation
- Community support, including gardening sessions and ongoing assistance for local estates
- Community improvements – repairing and renovating communities halls to be utilised by local residents
- Sustainability initiatives

OCO Ltd, Unit 4, Ashgrove Estate, Ashgrove Road, Bromley, Kent, BR1 4JW

020 8315 5600 • **slw@ocoltd.co.uk**

www.ocolimited.co.uk

PHYSICAL

Switches

- Hardware
- ££/£££
- Daily Living – Employment – Education – Social – Leisure

Switch access is a way of accessing technology using a simple switch, instead of complex interfaces such as keyboards or touch screens.

A switch is a simple device that has two states, on and off, just like a light switch. They come in different shapes and sizes and are operated in different ways. This is usually via a 'press' but you'll find switches that you grasp, pull, sip, puff or blink to activate. A common type is a buddy button or jelly bean switch.

Switches are used by people who have physical impairments and are unable to access a keyboard or mouse. Switches may also be suitable for some people with intellectual disabilities as they are a simple method of accessing games and other activities.

Primary switching involves using a switch to operate a computer, communication aid, environmental control or wheelchair. The user relies entirely on a switch or a combination of switches for control. Use of switches ranges from single-switch 'hit-and-happen' games for early users and those with severe cognitive difficulties, up to multiple-switch 'scanning' for those who want to use a computer to write emails, documents, surf the web, and for AAC (alternative and augmentative communication). Both iOS and Android platforms have switch access built-in and are ready to go.

A switch may play a supplemental access role, e.g. when a switch is used to replace a mouse button or mouse alternative. This can be a very efficient solution for someone who is able to move the pointer around the screen but not hold it steady while they press the button. You may need a switch interface device to plug your switch into your phone, tablet or computer.

bltt.org/introduction-to-switch-access

www.inclusive.co.uk/hardware/switches-and-switch-mountings

www.pretorianuk.com/assistive-switches

COMMUNICATION

SymWriter 2

- Software
- FREE trial then £££
- Education

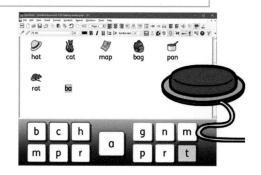

SymWriter 2 is a symbol word processor that allows users to see the meanings of words as they type.

Symbol supported documents visually assist the writer to process their writing. It is a useful classroom tool for teachers working with students who face cognitive and/or physical challenges to writing.

SymWriter 2 uses Widgit Symbols representing 40,000 English words. They can be used to create classroom materials and empower students to record and present their knowledge.

The program is easily customisable to support the learning level and needs of an individual user.

Free updates

Available at: www.widgit.com

| VISION PHYSICAL COGNITIVE |

Talking Desktop Calculator – A4

- Hardware/Consumer tech
- ££
- Daily Living – Employment – Education

The talking A4-sized desktop calculator is a basic calculator. It is A4 size with large buttons and a clear display.

Function keys include square root, percentage, and memory keys. Other features include; selectable unit or digit spoken results, repeat function of last spoken amount, adjustable volume, and an earphone connection that allows for private use when connected.

Other large calculators are also available without the speech function and can be helpful for people with low vision or physical needs.

Available at: www.amazon.co.uk

| HEARING COMMUNICATION COGNITIVE AUTISM |

TapSOS

- App (mobile)
- FREE
- Daily living

TapSOS is a smartphone app that provides users with a non-verbal method to connect with emergency services.

This could be useful for users who have hearing impairments, physical barriers to speech or cognitive disabilities such as autism.

When users have installed the app, they are asked to sign up and create a username and password.

They will also complete a personal profile with their name, email address, phone number, etc. and will be instructed to register with the BT eSMS service from within the app

In an emergency, a user can log into the app and select an icon for which emergency service is needed.

The phone's GPS pinpoints and reports the user's exact location. The user will be asked to provide additional information about the incident.

Depending on their selection of icons and responses to questions, TapSOS will create an SMS on their behalf to forward to the Emergency Services. A 999 call handler will then take over and organise an appropriate response.

If the call handler needs to follow up with the user or ask further questions, this will be done via SMS.

Available at: www.tapsos.com

VISION

TapTapSee

- App (mobile)
- FREE
- Daily Living – Employment – Education – Social – Leisure

Aimed at users with a range of visual impairments, TapTapSee is an app that helps them to easily identify any object.

A double-tap on the screen will photograph the object in question. The app will identify the object and speak aloud that identification.

The TapTapSee app can be a useful tool to help users with visual impairment become more independent in their daily lives. For example, TapTapSee could be used to identify food packets and medicine bottles or read the settings on a digital thermostat.

Please note: Users must have Talkback turned on for the spoken identification to work.

Features include photos can be taken from any angle, auto-focus notification, identify images from your camera roll, repeat last image identification, on/off flash toggle, share identification via text, email or social media.

Available at: taptapseeapp.com

COGNITIVE

TechSilver Dementia Tracker Keyring/Necklace

- Hardware/App (mobile)
- Cost of Tracker £££ Monthly Subscription £
- Daily living

The TechSilver Dementia Tracker Keyring/Necklace is a discreet wearable tracking device.

It can help a relative, friend or carer monitor the movements of a user with dementia or other cognitive impairment. It is also useful for tracking users with medical conditions that may put them at risk of seizures, falls, etc.

The TechSilver Dementia Tracker can be put in a bag, pocket or worn about the user's neck. The person monitoring the tracker can receive alerts via a free app on either an iPhone or Android smartphone. However, the alert service itself is subject to a monthly fee. Once paid, there is no limit to how many people monitor the tracker.

The TechSilver Dementia Tracker has an unlimited range and can be used anywhere in the UK or Europe.

Available at: www.techsilver.co.uk

VISION HEARING COMMUNICATION PHYSICAL COGNITIVE AUTISM DYSLEXIA

TechSilver Easy Tablet Computer

- Hardware/Consumer tech
- £££
- Daily Living – Employment – Education – Social – Leisure

TechSilver takes existing tablets and adapts them to be accessible to users with disabilities, particularly visual impairments.

They offer two models of Samsung tablet (the Samsung Galaxy Tab S610' and the Samsung Galaxy Tab A7 10.4-inch tablet). Their aim is to provide a device that offers the benefits of a standard tablet but has been customised to their needs. TechSilver offer personal tech support during the order process to help the user choose the best tablet for them.

The TechSilver Easy Tablet Computer menu is intuitive and easy to navigate. The tablet uses voice commands for hands-free control. Visual aspects such as colour schemes, contrast, and text and icon size are adjusted to suit the individual user.

A Tech Silver Easy Tablet Computer contains all the usual audiovisual and app-based features with additional features and tools to assist users with a visual impairment.

Available at: www.techsilver.co.uk

VISION

TechSilver Smartphone

- Consumer tech
- £££
- Daily Living – Employment – Education – Social – Leisure

TechSilver takes existing smartphones and customises them to be accessible to users with disabilities.

They offer two models of Samsung phone (the A32 and the lower-spec A22) adapted to be suitable for blind and visually impaired users.

A TechSilver adapted phone has easy to navigate menus that use high contrast colours and clear voice instructions to increase accessibility.

Communication adaptations include intuitive voice commands to control the phone and speech recognition tools for texts and messages. Pre-installed apps include a simplified version of Skype for video calls. There is a reading machine for printed text.

When ordering, users can request further customisation to suit their needs. On receiving the phone, the user also has access to a free one-hour set-up service.

Available at: www.techsilver.co.uk

COGNITIVE AUTISM DYSLEXIA

texthelp Read&Write

- Software
- £££
- Employment – Education

Read and Write is a literacy support tool that offers help with everyday tasks like reading text out loud, understanding unfamiliar words, researching Information, and proofing written work.

Some of the main features include
- Text-to-speech
- Text & Picture Dictionaries: Provide definitions and display images to help with word comprehension.
- Vocabulary List: Creates a list instantly into a new doc, including selected words, the dictionary definitions, images from Widgit Symbols, and an editable notes column.
- Check It: Reviews writing for incorrect grammar, spelling (phonetic), capitalisation, punctuation, verb tense and more.
- Audio Maker: Converts selected text into an audio file, and automatically downloads.
- Talk&Type: Turns the spoken word into text (unavailable for Read&Write for Mac users).

Text Help is widely used in education, especially for older learners. It is also very suitable in the workplace.

Available at: www.texthelp.com

VISION COGNITIVE AUTISM

Tile

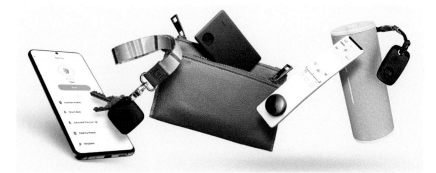

- Consumer tech
- ££/£££
- Daily Living – Employment – Education

Tile helps you keep track of your things. You can find misplaced things nearby and far away.

The Tile trackers and the free Tile app work with Android and Apple devices. The trackers connect via Bluetooth Low Energy (Bluetooth LE or BLE), allowing you to use a smartphone app to locate trackers. These are small tags that you attach to the item you want to be able to find, e.g. stuck to your phone, slipped into your wallet, attached to your keys.

Tile sells a range of different trackers – Pro, Mate, Slim and Sticker – which work with other Bluetooth-enabled devices, such as headphones.

You can search for items using the app on your phone or through Amazon Echo. The tag will start beeping when you ask for the Tile to be found, or you can search on a map if you have lost your belongings elsewhere.

Available at: uk.tile.com

COMMUNICATION

Tobii Dynavox Boardmaker 7

- Software
- £££
- Daily Living – Education – Social

Tobii Dynavox Boardmaker 7 software supports a user's learning, communication, and behavioural needs across various situations.

With access to over 50,000 readymade PCS (picture communication symbols), plus the option to create new symbols, an onscreen symbols board can be tailored to a user's individual needs.

Tobii Dynavox Boardmaker 7 software is intuitive and comes with built-in templates, editing capabilities, and advanced search options. It is also possible to import saved material from older Boardmaker versions.

Board Templates include options such as Activity Schedule, Matching, Sorting, Following Directions, Board Games, and many more.

All boards created can be printed for use away from the screen.

See also: Tobii Dynavox PCS Demonstration video: www.youtube.com/watch?v=J9OFF6D694I

Info at: uk.tobiidynavox.com

PHYSICAL COGNITIVE

Tobii Eye Tracker

- Software/hardware
- £££
- Daily Living – Employment – Education – Social – Leisure

Tobii Dynavox Eye Tracking software allows users to control their PC with their eyes. It is useful for users who experience physical barriers to using a standard keyboard or other manually operated computer controls.

The eye tracker emits near-infrared light, reflecting in the user's eyes. The eye tracker's camera picks up these reflections and uses them to calculate where the user is looking on the screen. Users can move to a specific area of the screen simply by moving the direction of their gaze.

Actions such as 'clicks' can be performed by the user blinking; by allowing their gaze to 'dwell' on a point for a set amount of time or by gazing at the screen while they click a physical switch.

Eye-tracking can be used to operate an online keyboard and access any functions available to users of a standard keyboard and mouse. It can assist users in writing documents and emails, making video calls, playing games, using the computer as an AAC (alternative and augmentative communication) device or as a remote control for home environmental technology.

Tobii Dynavox manufactures a range of built-in software apps and devices with eye-tracking software. For example, Windows Control enables computer access on a Windows PC, and TD Pilot is an eye-controlled communication device for an iPad.

Tobii Eye Tracker software does not currently work directly with Android devices. Using the Mirror for Android app on a Windows PC allows Android users to use eye-tracking to interact with their devices.

Info at: uk.tobiidynavox.com

COMMUNICATION COGNITIVE AUTISM DYSLEXIA

Tobii Dynavox Picture Communication Symbols

- Software
- Available on request
- Daily Living – Employment – Education – Social – Leisure

Tobii Dynavox Picture Communication Symbols (PCS) are symbol-based visual support designed to help people with learning, communication and behaviour challenges. They enable users to explore, understand and structure the world around them whilst allowing them to communicate their thoughts and express themselves more clearly.

Tobii Dynavox PCS is available in several styles depending on users' individual needs: PCS Classic, PCS ThinLine, PCS High Contrast, PCS InContext.

All styles of the PCS are available for licensing to anyone who wishes to use them.

Info at: uk.tobiidynavox.com

COGNITIVE AUTISM DYSLEXIA

ToDoist

 WEBSITE

- Software/App (mobile)
- FREE/££
- Daily Living – Employment – Education – Social

Todoist is an app that allows you to plan out your day and week. You can use the tool to add simple tasks that you tick off as you go and add descriptions for each of these.

To use Todoist, you can create a free account to get started. Alternatively, you can sign in with your Google account. The basic version is free, and this includes:
- 300 active tasks per project
- 20 active sections per project

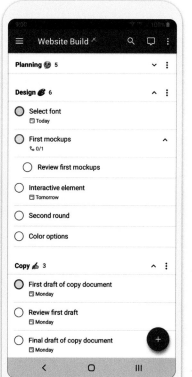

- A maximum of five collaborators on each project

You can upgrade to one of the paid plans if you want additional features.

You can use Todoist to:
- Capture and organise tasks the moment they pop into your head.
- Remember deadlines with reminders and due dates.
- Build lasting habits with recurring due dates like 'every Monday'.
- Collaborate on projects by assigning tasks to others.
- Prioritise your tasks with priority levels.
- Track your progress with personalised productivity trends.
- Integrate with Gmail, Google Calendar, Slack, Amazon Alexa and more.

ToDoist will allow you to enter your tasks in many ways including abbreviations and natural language.

Available at: todoist.com

PHYSICAL

Touchpads

- Hardware
- ££/£££
- Daily Living – Employment – Education

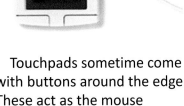

A touchpad is the small mouse pad built into almost every laptop.

Some people find the small movements required to operate these easier than the standard mouse. An example is the Cirque Smart Cat which is a touchpad that sits away from your body. It can be plugged into any computer and positioned easily so that small finger movements can control the position of the cursor on the screen.

The active area of the touchpad can be customised to make it possible to access everywhere on the screen, even with the movement of a single finger.

Touchpads sometime come with buttons around the edge. These act as the mouse buttons, although in most cases a simple tap of a finger on the pad can also be used instead of a mouse click.

Touchpads can be found at most computer shops and a number are available on Amazon.

PHYSICAL

Track-IT!

     WEBSITE

- Hardware
- £££
- Employment – Education – Leisure

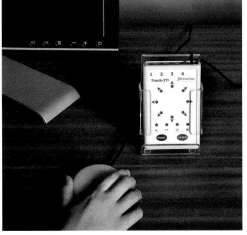

Track-IT! Is a remote scanning mouse suitable for users with motor skill difficulties who find it challenging to control a standard computer mouse.

The Track-IT houses a display of eight LED lights in a clock-like pattern. In the default single switch setting, the lights travel around clockwise, each light indicating a cursor movement or a switch operation.

When the light for the required function is reached, the user presses the switch to halt the scanning. A second press of the switch carries out the operation. (After a period of non-use, the Track-IT will re-commence scanning.)

The Track-IT has four sockets. As described, it can be operated as a single switch that allows complete control of the cursor and mouse button clicks. But for users with more dexterity, a second switch can be added, and the scanning is performed manually with the first switch.

Track-IT! has options to control the direction and speed at which the lights move, etc. These can be programmed, allowing the unit to be personalised according to a user's needs.

Available at: www.pretorianuk.com

PHYSICAL

Trackballs and rollerballs

- Hardware
- ££/£££
- Daily Living – Employment – Education – Leisure

A trackball or rollerball can be thought of as an old-fashioned mouse that has been turned upside down.

Rather than rolling the ball across a surface, you roll the ball with your fingers to control the cursor.

Trackballs are helpful as they don't need the dexterity and control that is needed to use a standard mouse. Because the buttons are separate from the ball, it is easier to hold the cursor over an item onscreen while clicking.

A good example is the BigTrack. This is a sturdy rollerball designed for early learners. It has a large, bright yellow ball that rolls smoothly and two large buttons as the buttons of a mouse. It doesn't require fine finger control and the ball is robust enough to be moved (and the buttons pushed) using palms, forearms or feet.

If you have problems pressing the buttons you can attach switches to the BigTrack.

The switches can be activated by another part of your body when you want to select an onscreen item.

Trackballs and rollerballs also make good game controllers when a standard mouse or joystick is difficult to use.

Most computer stores sell some forms of trackball and many can be found on amazon and mainstream suppliers. Specialist trackballs of different sizes and shapes are available for most AT suppliers.

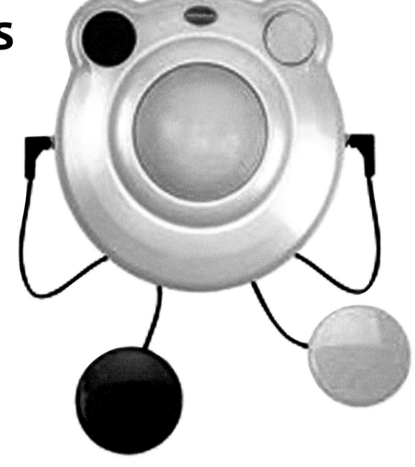

PHYSICAL

TrackerPro 2

   WEBSITE

- Hardware
- £££
- Daily Living – Employment – Education – Leisure

TrackerPro 2 is a head mouse designed for users who face challenges using a traditional computer mouse. It is completely hands-free and is operated using a small reflective dot affixed to the user's forehead, nose or glasses. When the user moves their head, the dot's movement is tracked by a camera unit mounted on the monitor.

The user's head movements are translated into cursor movements on the screen. Mouse clicks can be achieved by adding switches for physical clicking or Dwell selection software for automatic clicking. (Switches/Dwell software must be purchased separately).

Demonstration Video: www.youtube.com/watch?v=w44iMoop368

Info at: www.ablenetinc.com

PHYSICAL

TrackIR

- Software/Hardware/ Consumer tech
- ££/£££
- Leisure

TrackIR is an optical motion tracking game controller for use with Microsoft Windows.

It allows hands-free control when playing video games. Users wear a head tracking input device that transmits invisible infrared rays. A video camera mounted on top of the users' monitor observes these infrared rays translates them into movement on the screen.

The TrackIR tracks head movement in six different directions in 3D space. Each direction is on a separate axis. A user's actual movements are accurately linked to the in-game view.

Every purchase of a TrackIR camera comes with a TrackClip transmitter which attaches to a cap or visor worn by the user. In addition, it is possible to purchase a TrackClip Pro, which attaches

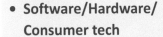

to most PC headsets and gives the added benefit of being able to track from a further distance away.

As well as purchasing the equipment, users will also need to download TrackIR software.

Available at: www.naturalpoint.com

HEARING COMMUNICATION

UbiDuo 3

- Hardware
- £££
- Daily Living – Employment – Education – Social

The UbiDuo 3 communication device is designed for users with hearing impairments.

It is a split-screen device that allows real-time conversation through typed messages. Unlike phone texting, there are no delays waiting for the message to arrive. Instead, it appears instantly on the linked screen.

The UbiDuo 3 comes in both wired and wireless versions. It has two 7-inch touchscreens and a long-lasting rechargeable battery allowing for up to eight hours of conversation.

The UbiDuo's 4GB storage enables users to store conversations. If extra capacity is needed, there is a USB port to connect a flash drive. If required, the device can be connected to a television or projector via HDMI to share the user's ideas with a wider audience.

The text and background colours are fully customisable to suit individual needs and preferences, with font sizes ranging from 12-72pt.

Available at: scomm.com

PHYSICAL

Upright Go 2

- Consumer tech
- FREE/£/££/£££
- Daily living

The upright Go 2 is a posture corrector and trainer. It is a small, light, comfortable strapless back posture corrector that you place on your upper back with an easy-to-use adhesive.

It reminds you to sit or stand up straight with a real-time gentle vibration reminder. The app syncs to your smartphone with 1 touch and generates a personalised training program of daily goals to improve your posture. These daily goals will help strengthen your back and train your brain to be aware of Slouching. Long-term habit formation for a healthy body

The wearable device attaches directly to your back, using 6 precision-enhanced sensors for the most exact measurements and feedback. You can use the app for tracking and monitoring your posture and get a daily score and see your progress over time.

Available at: www.uprightpose.com

HEARING

VC10 Vibrating Analogue Alarm Clock

- Hardware
- ££
- Daily living

The VC10 vibrating analogue clock looks like a conventional battery-powered analogue alarm clock.

What makes it different is the addition of a battery-powered vibrating pad which can be popped under the user's pillow to provide a physical alert when the alarm activates.

The clock has a choice of alarm settings, buzz, vibrate or both. If the alarm is not switched off, the buzzer and the vibration gradually increase in both volume and frequency. However, there is a snooze button that gives approximately a 4-minute respite. The snooze button also serves as a switch to temporarily illuminate the clock face.

The VC10 vibrating analogue alarm clock is a useful tool for users with hearing impairments.

Retailers include:
www.connevans.co.uk
www.cfdshop.org.uk
www.humantechnik.com

VISION DYSLEXIA

Victor Reader Stream

- Hardware
- £££
- Daily Living – Employment – Education – Social – Leisure

The Victor Reader Stream is a handheld media player. Users can listen to newspapers, books, internet radio, music, and other online resources.

Its numeric keypad has specific key shapes for different functions. Users have access to a wide range of online libraries in various languages, including DAISY libraries. It is also possible to transfer books in from systems such as Audible.

Although an audio device, the Victor Reader Stream has many navigation features that make it easy to move around a text or book. For example:

- users can browse books by chapter, section, page or phrase
- Information key – states book title, the time elapsed/remaining, total pages, etc
- Go To Page key – navigates to a specific page. (Users can then bookmark the page and even add a recording of their own voice annotations)
- Where Am I key – states reading position such as page, chapter or song title
- Sleep key – stops playback if the user falls asleep (then later, the VR Stream automatically resumes from the right place)
- Key Describer – describes the functions of each key

The Victor Reader Stream is a useful tool for users with a visual impairment or a reading disorder such as dyslexia.

Available at: store.humanware.com

COMMUNICATION COGNITIVE AUTISM

Visual Schedule Planner

- App (mobile)
- ££
- Daily Living – Employment – Education – Social

Visual Schedule Planner is a customisable visual schedule app that is designed to give you an audio/visual representation of events in the day.

Events that require more support can be linked to an 'activity schedule'. It is helpful in most situations and is designed for people who find visual support helps ease transitions and anxiety or who need a way to visually represent their day to help with memory and planning. Key features include viewing events daily, weekly or via a monthly calendar, using custom images, building activity schedules, timers, checklists, reminders and notes

Visual Schedule Planner has an easy-to-use interface, which can be adapted to meet your needs and preferences.

The product may no longer be developed but it remains available on Apple's App Store.

Available at: apps.apple.com

PHYSICAL COGNITIVE

VoiceAttack

- Software
- £
- Leisure

VoiceAttack allows users to add their voice as a controller to their games and apps.

They can create their own voice commands specific to the game or app they are using. It is a useful tool for users with physical disabilities that create barriers to using a keyboard.

VoiceAttack can be used to create macros that replace keystrokes/keystroke combinations or mouse clicks.

As well as game commands, the user can also program phrases such as 'Launch Spotify' to have control over their apps.

Available at: voiceattack.com

VISION COMMUNICATION PHYSICAL COGNITIVE AUTISM DYSLEXIA

Voice Record Pro

- APP (mobile)
- FREE
- Daily Living – Employment – Education – Social

Voice Record Pro is a professional voice recorder. It allows you to record voice memos, lectures, notes or discussions at unlimited length with configurable quality. Voice Record Pro can record directly as an MP4, MP3 or WAV file and has a convert function for all supported formats.

The recorder files can be:
- Exported or imported to different cloud storage space
- Sent to other devices via Bluetooth
- Downloaded directly to PC
- Posted as a Movie Clip on YouTube
- Exported to other apps on your device that can handle the audio file
- Imported from other apps
- Sent by email or message

You can also add notes, add a photo, add bookmarks, split or join recordings

Apply effects to the file including Echo, Volume, Pitch, and Speed.

Available at: apps.apple.com

VISION HEARING COMMUNICATION PHYSICAL COGNITIVE AUTISM DYSLEXIA

Voice Recorder

- App (mobile)
- FREE
- Daily Living – Employment – Education – Social

Voice Recorder is a free app that allows you to record your meetings, lectures or notes. There are restrictions on the length of recording.

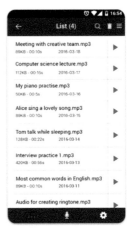

Features include
- High quality recording
- Simple user interface, easy to use.
- Support for mp3 files
- Play, pause, stop recording and playback
- Send/Share your recording.
- Delete your recording right from the app.
- Save the recording file.
- Background recording when the display is off.

Available at: play.google.com

COMMUNICATION

VoiceITT

- App (mobile)
- FREE
- Daily Living – Employment – Education – Social – Leisure

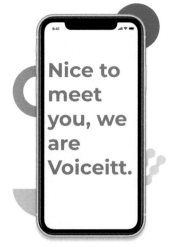

The VoiceITT app advertises itself as providing 'Speech recognition for all!'

The app is designed to understand non-standard speech and then repeat it aloud in a voice more easily understood by the average listener. The aim is to give a voice to people experiencing barriers to speech, including those with specific speech disabilities or whose speech is impeded by conditions such as Cerebral Palsy, Parkinson's Disease, ALS, and many more.

A user can train the app to recognise their distinct voice, pronunciation, speech pattern, etc., and link it to a specific phrase. Initially, a user may wish to choose from the 250+ everyday phrases already in the VoiceITT app, but there is also the option for them to add their own phrases as required. Training usually consists of five repetitions of each word. VoiceITT can recognise any speech, so it can be used by speakers of any language, with any accent.

Once the user has trained the app, they can use their voice to trigger VoiceITT to predict and speak aloud the corresponding phase. The more a user uses VoiceITT, the better it gets at predicting the phrase they want.

VoiceITT can communicate with other people with phrases such as "Hello" or "Nice to meet you." Or it can issue voice-controlled commands to a user's smart home devices, e.g. "Turn on the TV."

Available at: voiceitt.com

VISION HEARING PHYSICAL COGNITIVE AUTISM DYSLEXIA

Waze

 WEBSITE

- App (mobile)/Consumer tech
- FREE
- Daily Living – Employment – Education – Social – Leisure

Waze is an example of satellite navigation software. It works on smartphones and other computers that have GPS support.

It provides turn-by-turn navigation information and user-submitted travel times and route details while downloading local information over your internet connection

It is community-driven using the data of every user to inform other users of travel issues around them. The app creates revenue from very local advertising as you travel.

Waze has the ability to direct you on routes based on crowdsourced information. Waze users are able to report traffic-related incidents from accidents to police traps. This data is used by Waze to help other users either by alerting them of the condition ahead or rerouting the user to avoid the area entirely.

In addition to user input, Waze relies on information from public agencies for traffic events such as road construction. The more people that provide data the more accurate it becomes.

Available at: www.waze.com

VISION HEARING PHYSICAL COGNITIVE

WelcoMe by Neatebox

- App (mobile)
- FREE
- Daily Living – Social – Leisure

The WelcoMe by Neatebox mobile app is intended for users with various disabilities. It aims to address barriers to 'receiving a level of customer service that others take for granted'.

The app provides a service to pre-advise venues of any special needs a customer may have so their experience may go as smoothly as possible.

Welcome can be used to discover nearby participating venues and create a booking to notify them of a user's intended visit. Completing a tick box list informs the venue of the specific requirements. For example, if they require wheelchair access, have a hearing impairment or will have a guide dog with them. The user then receives an in-

app confirmation that the venue has been notified.

The app can only work with pre-arranged venues, but there is the facility to request a particular venue be added to the scheme.

Available at: www.neatebox.com

PHYSICAL

Wheelchair Calorimeter

- App (mobile)
- FREE
- Daily Living

The app uses your GPS location services from your iPhone to help you determine how many Calories you burn while exercising in a manual wheelchair.

It is designed for outdoor exercise. It is simple to use, you set your weight, press

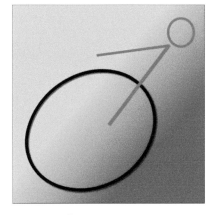

start, and start monitoring.

You can customise entries with your weight, as well as the weight of your

wheelchair/hand-bike and other equipment.

The app keeps an editable log of your exercise sessions, denoting distance, uphill climb, and calories burned in either metric or standard units of measurement.

The App can run in the background so you can set your music, send a text message or do anything else on your iPhone without having to stop or pause your exercise session.

Wheelchair Calorimeter will display a bar at the top of your home screen if it is running in the background so that you always know if it's running.

Available at: apps.apple.com

222

> PHYSICAL

Wheelmap

 WEBSITE

- Software/Hardware/ App (mobile)/Content/ Consumer tech
- FREE
- Daily Living – Social – Leisure

Using Wheelmap you can find and rate wheelchair-accessible locations across the world free of charge.

Wheelmap will help you to find wheelchair-accessible restaurants, cafes, toilets, shops, cinemas, parking lots, bus stops, and others.

The users have rated about one million places and continue to grow.

You can contribute information about places worldwide, adding your own experience and informing others.

For instance, you can rate the entrances and facilities of public places for their accessibility and upload images of the places to add extra information.

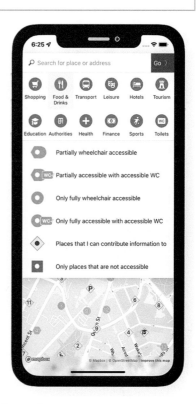

Available at: wheelmap.org

> PHYSICAL

Which Way Out

- App (mobile)
- £
- Education – Social – Leisure

Which Way Out is a maze/puzzle game designed for users to practice using single switch scan controls or step switch controls to move in different directions.

The game consists of 20 levels of increasing difficulty. Each level generates a random maze.

As players progress through

the levels, extra difficulty is introduced by introducing a dragon who will catch players who make poor choices or navigate the labyrinth too slowly.

The game is designed to be operable either by the

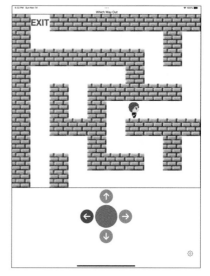

onscreen touch controls or by connecting it to an external control such as a keyboard or external four-step switches.

Auto-scan and Bluetooth external switch controls are included in the app as standard.

Available at: www.computerade.com

VISION DYSLEXIA

Windows High Contrast Settings

- Integrated
- FREE
- Daily Living – Employment – Education

Windows High Contrast Settings allows a user to change the contrast and colours on the screen.

Some colour combinations or low contrast can make text very difficult to read if a user has a visual impairment or a barrier to reading, such as dyslexia. Using Windows High Contrast Settings can make text easier and more comfortable to read.

If users find that none of the default themes fully fit their needs, they can customise the settings to suit. For example, if a user has a form of colour blindness, they can adjust the settings to take out the affected colours

The settings can modify the colour of a range of elements in Windows, such as the text and its background, selected or disabled (greyed out) text, text on buttons and hyperlinks.

Info at: support.microsoft.com

VISION PHYSICAL DYSLEXIA

Windows Keyboard Settings

- Integrated
- FREE
- Daily Living – Employment – Education – Social – Leisure

Windows Keyboard Settings office the user a variety of ways to customise the keyboard to make it easier and more comfortable to use.

Three of these options can be particularly beneficial for people who face physical challenges using a standard keyboard. They are the option to have an onscreen keyboard, 'sticky keys' and 'filter keys'.

Using an onscreen keyboard means that instead of physically pressing keys, users can use their mouse to select a graphical key on the screen. This could be teamed with a voice-controlled mouse cursor.

When Sticky Keys are turned on, the keyboard acts as if the key is being held down for longer. When a user is carrying out a multi-key command, such as Ctrl + Alt + Delete, the keys do not need to be pressed simultaneously but can be selected one after the other. This is a useful feature for users with poor manual dexterity.

When using the Filter Keys feature, Windows will monitor for repeated keystrokes and avoid a string of repeated characters (e.g. k rather kkkkkkkkkkk). This can help make typing less frustrating for users with hand tremors or other hand control limitations.

Info at: support.microsoft.com

VISION

Windows Magnifier

- Integrated
- FREE
- Daily Living – Employment – Education

The Windows Magnifier is integrated into Windows software, with no need for users to download or install it separately. It will magnify the screen up to 1600% and is a useful tool for users with visual impairments.

Once launched, the Windows Magnifier offers three different views: Full Screen, Lens and Docked.

As the name suggests, Full Screen magnifies the entire screen. Lens view magnifies only a small selection of the screen immediately around the mouse pointer.

In Docked view, the screen is horizontally divided into two. The upper half is a magnified version of the lower half.

The Windows Magnifier and its onscreen toolbar can be accessed and controlled from the user's keyboard or touchscreen. It allows the user to change the magnification level, smooth the edges of the magnified text, and invert the contrast/colours.

Info at: support.microsoft.com

VISION PHYSICAL COGNITIVE DYSLEXIA

Windows Mouse Options

- Integrated
- FREE
- Daily Living – Employment – Education – Social – Leisure

Windows Mouse Options allow users to customise the mouse pointer or cursor to suit their individual preferences. Some of the options regulate how the cursor operates (e.g. whether the left or right mouse button is the primary control).

Others, however, allow for adjustments to be made as to how the mouse pointer appears onscreen. These are of particular benefit to users who have a range of visual impairments. They can help them see and track the mouse pointer more easily.

Some of these visual options address the appearance of the mouse pointer, such as colour and size. Others affect the way it moves. It is possible to adjust the speed and which the pointer moves across the screen and to opt to have the pointer leave a trail behind it.

There is also the option to turn on a feature to show the mouse pointer's location whenever the Ctrl key is pressed. It works by placing a circle around the pointer, which quickly contracts, centering on the pointer, and then disappears. This is useful for finding the cursor on the screen is difficult.

Info at: support.microsoft.com

225

VISION DYSLEXIA

Windows Narrator

- Integrated
- FREE
- Daily Living – Employment – Education

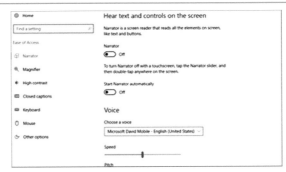

The Windows Narrator screen reader is integrated into Windows software with no need for users to download or install it separately.

It can be used to audibly 'narrate 'the user's actions as they navigate around the screen. For example, it will read aloud menus and the user's selections.

Narrator supports a variety of online activities. In Outlook, it can be used to read an email's status. When the email is opened, it will automatically read it aloud. When the user hovers over a link, Narrator will read the title of the webpage and then start to read the contents on the page as soon as it is opened.

When needed, Narrator will announce information about formatting, punctuation, font, etc.

Narrator supports a wide variety of refreshable braille displays, providing they connect via USB or a serial port. It is customisable to suit a user's preference in terms of voice, volume and rate of speech, and verbosity.

Info at: support.microsoft.com

VISION COMMUNICATION PHYSICAL COGNITIVE AUTISM DYSLEXIA

Windows Speech Recognition

- Integrated
- FREE
- Daily Living – Employment – Education – Social – Leisure

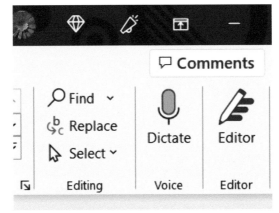

Recognition tool is included as standard with Windows software. It works with a small selection of languages (Including English, Spanish and Mandarin).

It can be used for text dictation and as a voice control tool to replace many keyboard commands and mouse functions.

When used for text dictation, Speech Recognition can be toggled on and off by clicking on the microphone icon in the toolbar. As well as dictating words, users can issue punctuation and other instructions such as full stop, comma, new line, go to end of the paragraph, select word, etc. Users can add words to the Speech Recognition dictionary and record their own pronunciations to increase recognition accuracy.

Windows Speech Recognition users have access to various voice commands to control their computers. For example, to open the start screen, select/open/close items, apps and files, and scroll the screen.

Info at: support.microsoft.com

HEARING

Wireless sound streaming systems

 WEBSITE

- Hardware
- £££
- Daily Living – Employment – Education – Social – Leisure

If someone has a hearing impairment, they may find a wireless sound streaming system can come in useful.

Such systems normally consist of a transmitter (usually with a built-in microphone) and a receiver. The transmitter captures the sound and sends it to the receiver. This can then relay it to the user via a listening device such as headphones/earphones or, in some cases, directly to their personal hearing aids.

Wireless sound streaming systems can be used in a variety of situations. If the user is having a face-to-face chat at home or perhaps in a pub or café, angling the transmitter towards the other speakers, the conversation can be picked up and transmitted to the user to hear it clearly without the distraction of background sounds.

Alternatively, it can be useful in a work or educational environment where if the speaker may not be face-to-face with the user but needs to hear them speak from some distance away. In a lecture hall, for example. The transmitter can be placed close to the speaker in those situations, whose words will then be transmitted to the user.

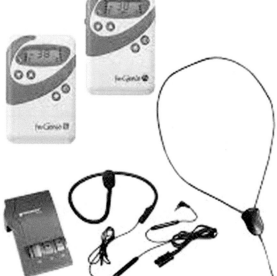

Wireless sound streaming systems can also listen to music or television. Many systems can be connected to a TV set or multimedia player via Bluetooth or perhaps plugged into the device's audio jack.

This means that users can more fully enjoy music and television shows etc, when by themselves. Still, if watching in company, they can listen through their hearing devices without increasing the volume for everybody else.

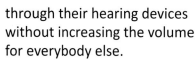

There are many different brands of wireless sound systems, such as Oticon, Phonak, and Bellman.

Some come with listening aids included as part of a package. Some are designed to be paired with the user's own listening aids.

MENTAL HEALTH

Wisdo

- App (mobile)
- Free or Monthly subscription = £
- Daily living

Wisdo: Experience Happier Days is a mental-health and wellbeing app that seeks to provide an alternative to the existing mainstream social networks.

Its aim is to connect users who have had similar experiences and so can offer each other non-judgemental support, understanding and encouragement.

New users are invited to check in to share their stories and long-term goals. They can then be connected to other users with similar life circumstances. At the same time, the app tracks their progress towards achieving goals and provides daily tips for success.

Wisdo hosts over 70 demographic-specific communities offering 24/7 support for issues such as self-esteem, addiction, PTSD, racism and many more.

The app also offers a Life Coach platform with 30+ professionally trained life coaches available to support users in achieving their goals. In addition, trained Wisdo Helpers and Guides are on hand to help users navigate the app.

Available at: wisdo.com

PHYSICAL

Withings BPM Connect Wi-Fi Smart Blood Pressure Monitor

- Hardware/App (mobile)
- ££/App = FREE
- Daily Living

The Withings BPM Connect Wi-Fi Smart Blood Pressure Monitor has large LED readings and colour-coded indicators that make it easy to understand blood pressure numbers.

Use it in conjunction with the Withings Health Mate app to record all measurements and link with other health apps, such as Google Fit, to easily track all a user's health data.

The Withings Health Mate app can highlight and interpret a user's health data. It compiles information that can be shared with healthcare professionals such as an ECG record or blood pressure data. Users can access a full health report shared with a medical practitioner via PDF.

App features include:
- Weight and body composition monitoring.
- Activity/sport monitoring.
- Sleep analysis/sleep apnoea detection.
- Hypertension management.
- Cardiovascular disease detection.

Available at: www.withings.com

PHYSICAL

Withings Thermo Smart Temporal Thermometer

- Hardware/App (mobile)
- ££/App = FREE
- Daily living

The Withings Thermo Smart Temporal Thermometer is a totally contact-free thermometer that can take and display a user's temperature in just two seconds.

Hold the thermometer 1 cm away from the forehead and move it across. The reading will instantly appear on the device, with the user's name and a colour-coded LED to tell you if the temperature is normal, elevated or high based on their age.

Using the app, you can set up users and create a record of their history such as temperature, symptoms, and medicine taken. There is also the option to add photos to display visual information,

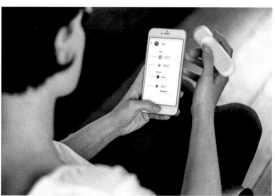

such as a picture of a rash.

The thermometer can be linked to up to eight users and can set reminders to take a temperature.

Available at: www.withings.com

MENTAL HEALTH

Worry Watch

- APP (mobile)/Content
- FREE or annual subscription = £
- Daily Living

The Worry Watch app aims to help its users manage their anxiety and moods. It offers guided anxiety journaling, coping techniques and mood recording and gives positive reinforcements.

The app encourages users to write down their thoughts

and feelings in the Anxiety Journal. It guides them through a process of Record → Reason → Respond → Reflect. Once users have used this process to identify triggers, the app leads them through guided coping techniques to help them respond to a stressful situation – delivered both visually and with a clear, soothing voice.

Users can record their moods in the Mood Journal as frequently. The app's Mood Tracker can help identify triggers and patterns associated with a user's changing moods and build positive moods.

Guided Positive Reinforcements include motivating quotes to promote positive thinking, notes to aid self-realisation and strategies for taking action.

NOTE: The 'Worry Watch' app is not intended to substitute or replace professional medical advice or treatment.

Available at: worrywatch.com

DYSLEXIA

XMind

- Software
- ££
- Employment – Education

XMind is a downloadable mind mapping software that converts your mind map into a slideshow. It can be used with a range of operating systems and devices.

XMind could benefit users who prefer to organise their work visually find it challenging to work with large amounts of text such as users with dyslexia.

Features include:
- Wide choice of colour themes/skeleton frameworks that can be combined according to the user's preferences.
- Choice of mind map styles (For example, Matrix, Fishbone, Timeline).
- Option to add shapes.
- Option to add maths equations.
- Image customisation.
- Auto-generated animated transmissions.

Example of mind maps at: www.xmind.net/share

Available at: www.xmind.net

VISION HEARING COMMUNICATION PHYSICAL COGNITIVE AUTISM DYSLEXIA

Zoom

    WEBSITE

- Software/App (mobile)/ Consumer tech
- FREE/£/££
- Daily Living – Employment – Education – Social – Leisure

Zoom is one of the leading video conferencing software app, that allows you to interact with people when in-person meetings aren't possible, at work, in education, and for social activities.

Zoom has become an essential app for many people with a disability, helping them to break a sense of isolation and loneliness, especially during the Covid pandemic. It has also allowed many to continue their education and to keep working when travel has not been possible.

Zoom is a cloud-based video conferencing service you use to virtually meet with other people. These meetings can be video or audio-only or both. A Zoom Meeting refers to a video conferencing meeting that's hosted using Zoom which you can join with a computer tablet or phone. Zoom Rooms are usually created for sub groups of meetings and are available with the paid versions.

Zoom's main features include unlimited one-on-one meetings even with the free plan, screen sharing and recording your meetings or events for later review.

Available at: zoom.us

VISION
ZoomText

- Software
- £££
- Daily Living – Employment – Education – Social – Leisure

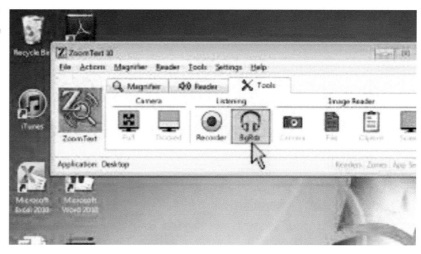

ZoomText Magnifier/Reader is a useful tool for users with low vision. It is a fully integrated magnification and reading program that enables users to enlarge and enhance everything on their computer screen. This makes information easier to see and use.

Features include:

- automatically reads documents, web pages, email, etc.
- echoes user's typing and essential programme activity
- up to 60x magnification
- range of colour schemes to reduce glare and maximise user's comfort
- a selection of highlighting features enables the user to monitor the position of the mouse pointer/text cursor, etc.
- supports MS touch-screen computers.

ZoomText online store currently only supports buyers in the USA. To purchase in the UK visit: www.sightandsound.co.uk

Info at: www.zoomtext.com • www.freedomscientific.com

liveit-project.eu

The LIVE-IT Project is an EU-funded survey on Cognitive Disability and Web Accessibility that conducted in-depth interviews, participant-led collaborative sessions, hackathons with high end-user involvement, and community-based testing of tools in four countries: Greece, Portugal, Ireland and the UK. It gathered data on the challenges associated with cognitive disabilities posed by web accessibility (which is ever-changing) and web-based tool use.

The LIVE-IT team developed nine categories of challenges: 1) Memory, 2) Focus, 3) Time and Organisation, 4) Mood, 5) Physical Barriers, 6) Reading, Writing and Comprehension, 7) Numbers, 8) Digital Skills, 9) General Problems. It mapped over 1000 web-based tools to those challenges through collaborative sessions with participants and hackathons. Some of those tools have spotlights in this Guide to Assistive Technology.

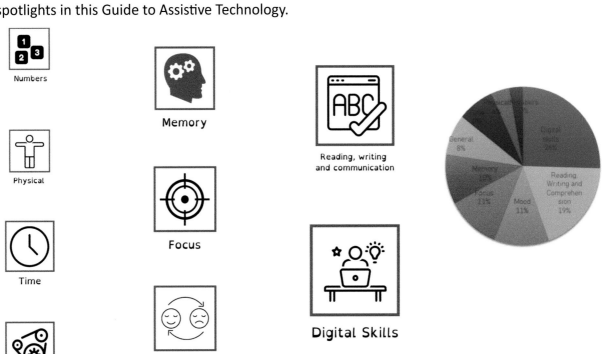

Toolkit

One of the major LIVE-IT Project outputs is the Open Toolkit, a dynamic web accessibility resource that was co-designed with participants. It has four sections: 1) a list of web accessibility contacts across Europe, 2) a catalogue of studies, policies, guidelines, and other resources, 3) an advisor tool to assist in the selection of web-based accessibility tools, 4) access to an online community dedicated to grassroots efforts to improve web accessibility. The Advisor Tool contains menus with the nine challenges and also sub-categories within each challenge. Individuals seeking out web-based tools to help with a specific challenge can use the menus to produce a list of tools that includes descriptions of each tool and information about how to access and use the tool (like what's provided in this book). Additionally, users can leave ratings and feedback on the tools for others to see when searching. To use the Open Toolkit:

live-it.azurewebsites.net/Toolkit/SearchToolkit

Hackathons

One of the major LIVE-IT Project activities was hosting a series of virtual hackathons with participants with cognitive disabilities from each of the four partner countries. The hackathons had themes: 1) authentication and online forms, 2) text-based and cluttered webpages, 3) interpreting information in various formats, 4) complex interfaces and tasks, 5) support and instructions. Each hackathon produced improvements for the Open Toolkit and improvements for existing tools. Results are at:

liveithackathon.wordpress.com

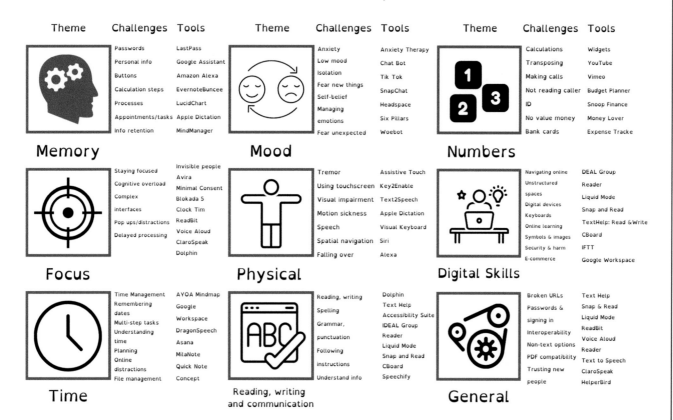

Other Projects and Organisations

The Easy Reading Project and the Insension Project are two EU-funded projects focused on digital or web accessibility. Their project websites are full of useful information for users, designers, educators, and researchers. AHEAD, an Irish non-profit, has created an online tool for finding assistive technology catered to specific needs. The University of Athens Department of Informatics hosts a searchable inventory of free assistive technology software for mobile devices called mATHENA, available in Greek and English.

Easy Reading Project – www.easyreading.eu/the-project

Insension Project – www.insension.eu

AHEAD AT Hive – ahead.ie/Discover-your-AT

mATHENA AT Inventory – athena.uoa.gr

Further sources of information

Organisations

Millennium Community Solutions – www.millenniumcommunitysolutions.com
AbilityNet – abilitynet.org.uk
Leonard Cheshire – www.leonardcheshire.org
Scope – www.scope.org.uk
Access to Work – www.gov.uk/access-to-work
Ace Centre – acecentre.org.uk

Publications

Assistive Technology Outcomes & Benefits (*online journal*) –
    www.atia.org/home/at-resources/atob

Online resources

Gari Database of Accessible Mobile Phones – gari.info
AI-driven tool for choosing AT – ATVisor.ai
Online Resource Directory – www.closingthegap.com/resource-directory
Catalogue of AT and technology for accessible education –
    accessibledigitallearning.org
AT resources and materials for education – includedu.online
Australian AT portal – www.atchat.com.au/your-portal
Resources for people with learning disabilities – develop.buddyproject.eu
Guide to adapting your technology to your needs – mcmw.abilitynet.org.uk

Newsletters

Access and Inclusion through Technology – www.accessandinclusion.news

Events

Naidex – www.naidex.co.uk
Sight Village – www.qac.ac.uk
ATIA (Assistive Technology Industry Association, US) – www.atia.org
BETT – www.bettshow.com

Product index by needs

AUTISM

74, 75, 76, 78, 79, 80, 81, 82, 85, 87, 92, 94, 95, 97, 99, 101, 106, 107, 114, 118, 119, 121, 122, 123, 124, 125, 128, 130, 134, 137, 139, 140, 141, 142, 143, 147, 149, 150, 151, 158, 160, 163, 164, 165, 167, 168, 171, 173, 174, 179, 180, 181, 184, 187, 191, 192, 193, 197, 200, 201, 202, 208, 210, 211, 213, 219, 220, 221, 226, 230, 235

COGNITIVE

74, 75, 76, 78, 79, 80, 81, 82, 85, 87, 91, 92, 93, 94, 95, 97, 99, 101, 104, 106, 107, 115, 118, 121, 122, 123, 124, 125, 128, 130, 134, 137, 139, 140, 141, 142, 143, 147, 149, 150, 151, 158, 163, 164, 165, 167, 171, 173, 174, 179, 180, 181, 184, 187, 192, 193, 197, 200, 201, 208, 209, 210, 211, 212, 213, 219, 220, 221, 222, 225, 226, 230, 235

COMMUNICATION

74, 75, 76, 78, 79, 80, 81, 82, 83, 85, 89, 92, 95, 97, 101, 109, 113, 114, 115, 118, 119, 122, 123, 124, 125, 128, 130, 135, 138, 139, 142, 143, 149, 150, 158, 159, 165, 171, 173, 174, 179, 180, 181, 183, 184, 185, 189, 191, 197, 200, 201, 202, 204, 207, 208, 210, 212, 213, 217, 219, 220, 221, 226, 230, 235

DYSLEXIA

74, 75, 76, 78, 79, 80, 81, †82, 87, 91, 92, 94, 95, 97, 101, 104, 106, 107, 111, 112, 113, 114, 115, 118, 119, 121, 122, 123, 124, 125, 128, 130, 134, 135, 137, 138, 140, 141, 142, 143, 145, 147, 149, 150, 151, 154, 158, 159, 160, 163, 164, 165, 167, 168, 171, 173, 174, 179, 180, 181, 184, 187, 192, 193, 197, 200, 210, 211, 213, 218, 220, 221, 224, 225, 226, 230, 235

HEARING

74, 75, 76, 77, 78, 79, 80, 81, 82, 85, 88, 89, 90, 92, 95, 97, 99, 100, 118, 121, 122, 123, 124, 125, 130, 134, 135, 143, 152, 158, 160, 165, 171, 173, 180, 181, 182, 184, 191, 193, 198,

200, 203, 204, 205, 208, 210, 217, 218, 220, 221, 222, 227, 230, 235

MENTAL HEALTH

98, 103, 106, 107, 109, 144, 147, 158, 192, 194, 197, 202, 228, 229, 235

PHYSICAL

18, 19, 74, 75, 76, 77, 78, 79, 80, 81, 82, 83, 84, 85, 87, 90, 91, 92, 93, 95, 99, 100, 101, 102, 104, 107, 108, 110, 111, 114, 115, 117, 118, 120, 121, 122, 123, 124, 125, 127, 129, 130, 131, 132, 133, 135, 137, 138, 141, 142, 143, 145, 148, 149, 150, 151, 153, 154, 155, 158, 159, 160, 161, 162, 163, 164, 165, 168, 169, 170, 171, 172, 173, 174, 175, 179, 180, 181, 184, 188, 189, 194, 197, 199, 200, 201, 202, 204, 207, 208, 210, 212, 214, 215, 216, 217, 219, 220, 221, 222, 223, 224, 225, 226, 228, 229, 230, 235

VISION

74, 75, 76, 78, 79, 80, 81, 82, 83, 84, 85, 87, 88, 91, 92, 93, 95, 98, 100, 101, 102, 103, 104, 108, 110, 111, 112, 113, 115, 118, 119, 120, 122, 123, 124, 125, 135, 140, 141, 143, 144, 145, 149, 151, 152, 154, 157, 158, 160, 161, 162, 165, 167, 169, 171, 172, 173, 174, 175, 177, 178, 179, 180, 181, 182, 183, 184, 187, 188, 190, 193, 195, 200, 203, 204, 205, 208, 209, 210, 211, 218, 220, 221, 222, 224, 225, 226, 230, 231, 235.

Product index by tasks and settings

Daily Living
74, 75, 76, 77, 78, 79, 81, 82, 83, 84, 85, 87, 88, 89, 90, 91, 92, 93, 94, 95, 97, 98, 99, 100, 101, 102, 103, 104, 106, 107, 108, 109, 110, 111, 113, 114, 115, 117, 118, 119, 120, 121, 122, 123, 124, 125, 127, 128, 129, 130, 131, 132, 134, 135, 137, 138, 139, 140, 141, 142, 143, 144, 145, 147, 148, 149, 150, 151, 152, 154, 155, 157, 158, 159, 160, 161, 162, 163, 165, 167, 168, 169, 170, 171, 172, 173, 174, 175, 178, 179, 180, 181, 182, 183, 184, 185, 187, 188, 189, 190, 191, 192, 193, 194, 195, 197, 198, 199, 200, 201, 202, 203, 204, 205, 207, 208, 209, 210, 211, 212, 213, 214, 215, 216, 217, 218, 219, 220, 221, 222, 223, 224, 225, 226, 227, 228, 229, 230, 231, 237

Education
76, 78, 79, 81, 82, 83, 84, 85, 87, 88, 89, 90, 91, 92, 94, 95, 97, 99, 100, 101, 102, 104, 106, 107, 109, 110, 111, 112, 113, 114, 115, 117, 118, 119, 120, 121, 122, 123, 124, 125, 127, 128, 130, 134, 135, 137, 138, 139, 140, 141, 142, 143, 144, 145, 148, 149, 150, 151, 152, 153, 154, 155, 158, 159, 160, 162, 163, 164, 165, 167, 168, 169, 172, 173, 174, 175, 177, 178, 179, 182, 183, 184, 185, 187, 188, 189, 190, 191, 192, 193, 194, 195, 197, 199, 200, 201, 202, 204, 205, 207, 208, 209, 210, 211, 212, 213, 214, 215, 216, 217, 218, 219, 220, 221, 223, 224, 225, 226, 227, 230, 231, 237

Employment
76, 78, 79, 81, 82, 83, 84, 85, 87, 88, 89, 90, 91, 92, 94, 95, 97, 99, 100, 101, 102, 104, 106, 107, 110, 111, 112, 113, 114, 115, 117, 118, 119, 120, 121, 122, 123, 124, 125, 127, 128, 134, 135, 137, 138, 139, 140, 141, 142, 143, 144, 145, 148, 149, 150, 151, 152, 153, 154, 155, 158, 159, 160, 162, 163, 164, 165, 167, 168, 169, 172, 173, 174, 175, 177, 178, 179, 182, 183, 184, 185, 187, 189, 190, 191, 192, 193, 194, 195, 197, 198, 199, 200, 201, 202, 204, 205, 207, 208, 209, 210, 211, 212, 213, 214, 215, 216, 217, 218, 219, 220, 221, 224, 225,

226, 227, 230, 231, 237

Leisure

74, 75, 76, 77, 78, 79, 80, 81, 82, 83, 84, 85, 87, 88, 90, 91, 92, 93, 94, 95, 97, 99, 101, 102, 103, 104, 106, 111, 113, 114, 115, 117, 118, 119, 120, 122, 123, 124, 125, 129, 131, 132, 133, 134, 137, 138, 139, 140, 141, 142, 143, 144, 145, 147, 148, 149, 150, 153, 154, 155, 157, 158, 159, 160, 161, 162, 163, 164, 165, 168, 169, 172, 173, 174, 175, 178, 179, 180, 181, 182, 183, 184, 185, 188, 189, 190, 191, 193, 194, 195, 197, 199, 200, 201, 204, 205, 207, 209, 210, 212, 213, 214, 215, 216, 218, 219, 221, 222, 223, 224, 225, 226, 227, 230, 231, 237

Social

74, 75, 76, 77, 78, 79, 82, 83, 84, 85, 87, 88, 89, 90, 91, 92, 93, 94, 95, 97, 99, 101, 102, 104, 106, 107, 109, 111, 113, 114, 115, 117, 118, 119, 121, 122, 123, 124, 125, 128, 134, 135, 137, 138, 139, 140, 141, 142, 143, 144, 145, 147, 148, 149, 150, 153, 154, 155, 158, 159, 160, 162, 163, 164, 165, 167, 168, 169, 172, 173, 174, 178, 179, 180, 181, 182, 183, 184, 185, 187, 189, 190, 191, 192, 193, 194, 195, 197, 199, 200, 201, 203, 204, 205, †206, 207, 209, 210, 212, 213, 217, 218, 219, 220, 221, 222, 223, 224, 225, 226, 227, 230, 231, 237.

Products that are free

74, 75, 76, 78, 79, 82, 83, 84, 85, 87, 88, 89, 90, 91, 92, 94, 95, 97, 98, 99, 100, 101, 103, 104, 107, 108, 109, 110, 111, 122, 123, 128, 129, 130, 132, 134, 135, 137, 138, 139, 140, 141, 142, 143, 144, 147, 148, 149, 158, 159, 160, 161, 162, 164, 167, 168, 169, 170, 171, 172, 174, 175, 179, 180, 181, 187, 188, 191, 192, 193, 194, 195, 197, 200, 202, 203, 204, 205, 207, 208, 209, 213, 217, 220, 221, 222, 223, 224, 225, 226, 228, 229, 230.

www.millenniumcommunitysolutions.com